Kelley Wingate
Math Practice

Fourth Grade

Credits
Content Editor: Amy R. Gamble
Copy Editor: Elise Craver

Visit *carsondellosa.com* for correlations to Common Core, state, national, and Canadian provincial standards.

Carson-Dellosa Publishing, LLC
PO Box 35665
Greensboro, NC 27425 USA
carsondellosa.com

ISBN 978-1-4838-0502-3
05-313151151

Table of Contents

Introduction

Competency in basic math skills creates a foundation for the successful use of math principles in the real world. Practicing math skills—in the areas of operations, algebra, place value, fractions, measurement, and geometry—is the best way to improve at them.

This book was developed to help students practice and master basic mathematical concepts. The practice pages can be used first to assess proficiency and later as basic skill practice. The extra practice will help students advance to more challenging math work with confidence. Help students catch up, stay up, and move ahead.

Common Core State Standards (CCSS) Alignment

This book supports standards-based instruction and is aligned to the CCSS. The standards are listed at the top of each page for easy reference. To help you meet instructional, remediation, and individualization goals, consult the Common Core State Standards alignment chart on page 4.

Leveled Activities

Instructional levels in this book vary. Each area of the book offers multilevel math activities so that learning can progress naturally. There are three levels, signified by one, two, or three dots at the bottom of the page:

- Level I: These activities will offer the most support.
- Level II: Some supportive measures are built in.
- Level III: Students will understand the concepts and be able to work independently.

All children learn at their own rate. Use your own judgment for introducing concepts to children when developmentally appropriate.

Hands-On Learning

Review is an important part of learning. It helps to ensure that skills are not only covered but are internalized. The flash cards at the back of this book will offer endless opportunities for review. Use them for a basic math facts drill, or to play bingo or other fun games.

There is also a certificate template at the back of this book for use as students excel at daily assignments or when they finish a unit.

Common Core State Standards Alignment Chart

Common Core State Standards*		Practice Page(s)
Operations and Algebraic Thinking		
Use the four operations with whole numbers to solve problems.	4.OA.1–4.OA.A.3	5–7
Gain familiarity with factors and multiples.	4.OA.4	8–10
Generate and analyze patterns.	4.OA.5	9
Number and Operations in Base Ten		
Generalize place value understanding for multi-digit whole numbers.	4.NBT.1–4.NBT.3	11–22
Use place value understanding and properties of operations to perform multi-digit arithmetic.	4.NBT.4–4.NBT.6	8–13, 23–37
Number and Operation—Fractions		
Extend understanding of fraction equivalence and ordering.	4.NF.1–4.NF.2	38–64
Build fractions from unit fractions by applying and extending previous understandings of operations on whole numbers.	4.NF.3–4.NF.4	53–67
Understand decimal notation for fractions, and compare decimal fractions.	4.NF.5–4.NF.7	68–73
Measurement and Data		
Solve problems involving measurement and conversion of measurements from a larger unit to a smaller unit.	4.MD.1–4.MD.3	74–85
Represent and interpret data.	4.MD.4	86–88
Geometric measurement: understand concepts of angle and measure angles.	4.MD.5–4.MD.7	89–94
Geometry		
Draw and identify lines and angles, and classify shapes by properties of their lines and angles.	4.G.1–4.G.3	92–103

Solving Word Problems

Circle the correct problem to solve each problem. Solve.

Last week, 415 people attended the school carnival. Of those, 316 rode the Ferris wheel. How many people did not ride the Ferris wheel?

99 people

$$\begin{array}{r} 415 \\ + \ 316 \end{array}$$

$$\begin{array}{r} 4\cancel{1}5 \\ - \ 316 \\ \hline 099 \end{array}$$

At the carnival, 236 people ate pink cotton candy, and 178 people ate blue cotton candy. How many people in all ate cotton candy?

414 people

$$\begin{array}{r} 236 \\ + \ 178 \\ \hline 414 \end{array}$$

$$\begin{array}{r} 236 \\ - \ 178 \end{array}$$

The carnival had 23 rides. If Whitney rode each ride 6 times, how many times did she ride in all?

138 rides

$$\begin{array}{r} 23 \\ + \ 6 \end{array}$$

$$\begin{array}{r} 23 \\ \times \ 6 \\ \hline 138 \end{array}$$

At the carnival, 160 kids played relay races. The kids were divided into 8 teams. How many kids were on each team?

20 kids

$$\begin{array}{r} 160 \\ \times \ 8 \end{array}$$

$$8\overline{)160}$$ with 20, -16, 000

Raul sold 594 ride tickets in 1 hour at the carnival. If he sells the same amount each hour, how many tickets will he sell in 4 hours?

2,376 tickets

$$\begin{array}{r} 594 \\ \times \ 4 \\ \hline 2376 \end{array}$$

$$4\overline{)594}$$

If 137 girls and 159 boys attended the carnival, how many more boys attended than girls?

22 more boys

$$\begin{array}{r} 137 \\ + \ 159 \end{array}$$

$$\begin{array}{r} 159 \\ - \ 137 \\ \hline 022 \end{array}$$

Solving Word Problems

Write one or more equations for each problem. Solve.

⭐1. The raffle ticket fund-raiser sold 2,453 tickets last year and 3,832 tickets ~~last~~ this year. How many more tickets did they sell this year than last year?

1,379 tickets

$$\begin{array}{r} 3,8\cancel{3}2 \\ - 2,453 \\ \hline 1,379 \end{array}$$

⭐2. Each student received 3 ticket books to sell raffle tickets. Each book had 50 tickets. If Ella turned in 98 unsold tickets, how many tickets did she sell?

52 tickets

$$3 \times 50 = \cancel{150} \\ \begin{array}{r} - 98 \\ \hline 52 \end{array}$$

⭐3. The bake sale fund-raiser sold five dozen chocolate chip cookies, nine dozen sugar cookies, and six dozen oatmeal cookies. How many cookies did the fund-raiser sell?

240

$$\begin{array}{r} 108 \\ + 72 \\ 60 \\ \hline 240 \end{array}$$

$$\begin{array}{r} 12 \\ \times 6 \\ \hline 72 \end{array}$$

$$\begin{array}{r} 12 \\ \times 9 \\ \hline 108 \end{array}$$

$$\begin{array}{r} 12 \\ \times 5 \\ \hline 60 \end{array}$$

⭐4. The bake sale made $832 on Friday and $1,276 on Saturday. How much money did the bake sale make in all?

$2108

$$\begin{array}{r} 1276 \\ + 832 \\ \hline 2108 \end{array}$$

⭐5. The school decided to divide the profits from their fund-raising between 9 classrooms in the school. They will put any leftover money toward a new welcome mat for the school. If the school raised a total of $4,749, how much money will go to each classroom? How much money will go to buy the new welcome mat?

$527 for each classroom and $6 gose to the to the welcome mat.

$$4749 \div 9 = 527 \text{ r } 6$$

⭐6. Kayla won the raffle for an afternoon at Ace Arcade. She gets 200 free tokens. If each game takes 3 tokens, how many games can she play? If she finds another token on the ground, can she play one more game? Explain.

$$\begin{array}{r} 66 \text{ r2} \\ 3 \overline{)200} \\ -18 \downarrow \\ \hline 020 \\ -18 \\ \hline 02 \end{array}$$

She can play 66 games. Yes she has enough to play 1 game.

Solving Word Problems

Solve each problem.

1. The fourth grade is going on a field trip to Colonial Town. Three fourth-grade classes are going, each with 19 students. One chaperone is needed for every 9 students. How many chaperones will need to go on the field trip?

    $57 \div 9 = 7$

    7 chaperones

    $\begin{array}{r} \overset{2}{1}9 \\ \times\ 3 \\ \hline 57 \text{ kids} \end{array}$

2. Colonial Town has an average of 7,895 total visitors on a weekend day and an average of 3,638 total visitors on a weekday. During the week, the average number of student visitors on field trips is 2,493. Not counting students on a field trip, how many more visitors on average are there on a weekend day than on a weekday?

    6750 more visitors

    $\begin{array}{r} 7895 \\ 1145 \\ \hline 6750 \end{array}$

    $\begin{array}{r} 3638 \\ -2493 \\ \hline 1145 \end{array}$

3. Students on a field trip to Colonial Town get to make their own candles. If the average number of students in a class is 23, and 38 classes of students have field trips each week, what is the average number of candles made by students each week?

    874 candles

    $\begin{array}{r} \overset{2}{3}8 \\ \times 23 \\ \hline 114 \\ +760 \\ \hline 874 \end{array}$

4. The teachers buy cookies from the bakery for the students. They want each of their 73 students to get 4 cookies. If the cookies come in packages of 9, how many packages do they need to buy?

    $\begin{array}{r} 32 \\ 9\overline{)292} \\ -27 \\ \hline 022 \end{array}$ $\begin{array}{r} 4\ 1\ 2\overline{)2} \\ -18 \\ \hline 04 \end{array}$

    33 packages

    $\begin{array}{r} 73 \\ \times 4 \\ \hline 292 \end{array}$

5. At the blacksmith's shop, the students learn that the blacksmith forge gets as hot as 1400°F. How many times hotter is the forge than the typical air temperature of 70°F?

    $70 \div 1400 = 20$ 20x hotter

    $\begin{array}{r} 20 \\ 70\overline{)1400} \\ -14 \\ \hline 000 \end{array}$

6. The blacksmith tells the students that he and his apprentice have been working on making nails for building projects and repairs in the town. They made 964 nails the first week of the month, 1,072 nails the second week, 936 nails the third week, and 1,113 nails the fourth week of the month. They will bundle the nails in boxes of 100. How many boxes will they need?

    41 boxes   4,085

    $1,072 + 1,113 + 964 + 936 = 4,085$

7. Write a division word problem in which you would have to ~~interpret the remainder.~~

    6 cookies 2 friends want cookies how many cookies well they get

Determining Factors and Multiples

Factors are numbers multiplied together. The first factor tells the number of sets. The second factor tells the number in each set.

How many different factors can you name for one number?

$2 \times 3 = 6$

$6 \times 1 = 6$

factors: 1, 2, 3, 6

Write all of the multiplication sentences for each set. Then, list the factors.

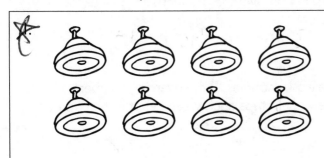

$4 \times 2 = 8$

$1 \times 8 = 8$

factors: 1, 2, 4, 8

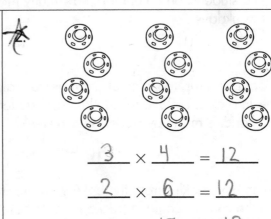

$3 \times 4 = 12$

$2 \times 6 = 12$

$1 \times 12 = 12$

factors: 1, 2, 3, 4, 6, 12

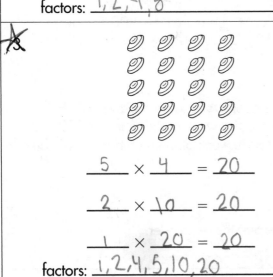

$5 \times 4 = 20$

$2 \times 10 = 20$

$1 \times 20 = 20$

factors: 1, 2, 4, 5, 10, 20

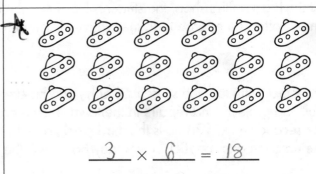

$3 \times 6 = 18$

$2 \times 9 = 18$

$1 \times 18 = 18$

factors: 1, 2, 3, 6, 9, 18

 What number, other than 1, is a common factor for all 4 numbers in problems 1–4? 2

So, we can say that all of these numbers are multiples of 2 .

Name _____

Determining Factors and Multiples

Multiples are numbers that all have the same particular factor.

Example: Even numbers are all multiples of the factor 2.

multiples:	2	4	6	8	10	12	14
other factor:	1	2	3	4	5	6	7

List the multiples of each factor. Write the other factor in the factor pair below each multiple.

1. multiples of the factor 5

multiples: 10 _15_ 20 _25_ _30_ 35 _40_ _45_

other factor: _1_ _2_ _3_ _4_ _5_ _6_ _7_ _8_

2. multiples of the factor 4

multiples: 12 _16_ 20 _24_ _28_ 32 _36_ 40

other factor: _2_ _3_ _4_ _5_ _6_ _7_ _8_ _9_

3. multiples of the factor 3

multiples: 15 _18_ _21_ 24 _27_ _30_ _33_ _36_

other factor: _5_ _6_ _7_ _8_ _9_ _10_ _11_ _12_

4. multiples of the factor 6

multiples: 6 12 18 _24_ _30_ 36 _42_ _48_

other factor: _1_ _2_ _3_ _4_ _5_ _6_ _7_ _8_

5. multiples of the factor 9

multiples: 18 _27_ _36_ _45_ 54 _63_ 72 _81_

other factor: _2_ _3_ _4_ _5_ _6_ _7_ _8_ _9_

6. multiples of the factor 8

multiples: 32 _40_ _48_ 56 _64_ _72_ 80 _88_

other factor: _4_ _5_ _6_ _7_ _8_ _9_ _10_ _11_

Determining Factors and Multiples

> A **composite number** has factors other than 1 and itself.
>
> A **prime number** has only the factors 1 and itself.

Name any factors, other than 1, that each group of composite numbers are multiples of.

1. 36, 27, 99, 45, 63: _1, 3, 9_

2. 40, 20, 70, 100, 60: _1, 2, 5, 10_

3. 8, 32, 12, 48, 20: _1, 2, 4_

4. 42, 21, 56, 84, 35: _1, 7_

5. 20, 10, 60, 40, 25: _1, 5_

Write 4 multiples that have both factors in common.

ex : 6. 2 and 3: _6, 12, 18, 24_

7. 4 and 6: _12, 24, 37, 48_

8. 5 and 2: _10, 20, 30, 40, 50, 60, 70, 80, 90, 100_

Write the factor pairs for each number. Tell whether it is a composite number or a prime number.

9. 36: _1, 2, 3, 6, 9, 36_ _____

10. 14: _1, 2, 7, 14_ _____

11. 23: _1, 23_ _____

12. 33: _1, 3, 11, 33_ _____

13. 48: _1, 2, 4, 6, 8, 48_ _____

14. 11: _1, 11_ _____

Using Place Value to Multiply and Divide

Multiplying by 10s adds places to a number.

$200 \times 10 = 2,000$ hundreds become thousands
$50 \times 100 = 5,000$ tens become thousands

Dividing by 10s removes places from a number.

$700 \div 10 = 70$ hundreds become tens
$70,000 \div 100 = 700$ ten thousands become hundreds

Use place value to multiply or divide.

1. $800 \times 10 =$ __8000__

2. $100,000 \div 10 =$ __10000__

3. $30 \times 10 =$ __300__

4. $10 \times 100 =$ __1000__

5. $9,000 \times 10 =$ __90,000__

6. $4 \times 100 =$ __400__

7. $50,000 \times 10 =$ __500,000__

8. $30 \times 100 =$ __3000__

9. $4,000 \div 10 =$ __400__

10. $8,000 \times 100 =$ __800,000__

11. $70 \div 10 =$ __70__

12. $200 \div 100 =$ __2__

13. $20,000 \div 10 =$ __2000__

14. $90,000 \div 100 =$ __900__

Using Place Value to Multiply and Divide

Remember: When you multiply by a multiple of 10, add the number of zeros in the multiple of 10 to the first number:

400 (2 zeros) × 100 (2 zeros) = 40,000 (4 zeros)

When you divide by a multiple of 10, subtract the number of zeros in the multiple of 10 from the first number:

40,000 (4 zeros) ÷ 100 (2 zeros) = 400 (2 zeros)

Use place value to multiply or divide.

1. 6,000 × 10 = 60,000

2. 60,000 ÷ 10 = 6,000

3. 70,000 × 10 = 700,000

4. 4,000 ÷ 100 = 40

5. 800 × 100 = 800,00

6. 700,000 ÷ 100 = 700

7. 2,000 × 100 = 200,000

8. 500 ÷ 10 = 50

9. 340 × 10 = 3400

10. 280,000 ÷ 10 = 28,00

11. 7,890 × 10 = 78,900

12. 79,000 ÷ 100 = 790

13. 459 × 1,000 = 459,000

14. 5,240 ÷ 10 = 524

15. 52 × 10,000 = 520,000

16. 582,000 ÷ 1,000 = 582

17. 4 × 100,000 = 400,000

18. 36,000 ÷ 1,000 = 36

19. 100,000 × 10 = 1,000,000

20. 900,000 ÷ 10,000 = 90

Using Place Value to Multiply and Divide

Use place value to multiply or divide.

1. $5 \times 70 =$ 350

2. $90 \times 2,000 =$ 180,000

3. $4 \times 9,000 =$ 36,000

4. $500 \div 5 =$ 100

5. $200,000 \times 3 =$ 600,000

6. $7,000 \div 70 =$ 100

7. $50 \times 400 =$ 20,000

8. $30,000 \div 300 =$ 100

9. $30,000 \times 30 =$ 900,000

10. $900 \div 90 =$ 100

11. $3 \times 800 =$ 24,000

12. $22,000 \div 2 =$ 11000

13. $9 \times 50,000 =$ 450,000

14. $450 \div 90 =$ 5

15. $600 \times 900 =$ 540000

16. $600 \div 30 =$ 20

17. $500 \times 300 =$ 150,000

18. $21,000 \div 7 =$ 3000

19. $40 \times 7,000 =$ 280,000

20. $800,000 \div 400 =$ 2000

21. $20 \times 40,000 =$ 800,000

22. $3,600 \div 90 =$ 40

Exploring Place Value   _100%_   ☺ ☆

> Numbers can be written in three ways.
>
> **Standard form** is a way to write a number that shows only the digits:
> 2,243
>
> **Expanded form** is a way to write a number that shows the place value of each digit:
> 2,000 + 200 + 40 + 3
>
> **Word form** is a way to write a number using number words:
> two thousand two hundred forty-three

Write each number in standard form.

1. 500 + 20 + 4 = _524_

2. 3,000 + 700 + 80 + 1 = _3781_

3. 60,000 + 1,000 + 900 + 30 + 2 = _61,932_

4. 800,000 + 90,000 + 5,000 + 400 + 10 + 6 = _895,416_

5. 400,000 + 70,000 + 3,000 + 200 + 60 = _473,260_

6. six hundred thirty-one = _631_

7. seven thousand four hundred twenty-five = _7,425_

8. ninety-three thousand eight hundred seventeen = _93,817_

9. one hundred twenty-one thousand three hundred seventy-six = _121,376_

10. forty-eight thousand one hundred sixty-nine = _48,169_

Exploring Place Value

> **Remember:**   **Standard form:**   35,894
>
> **Expanded form:**   30,000 + 5,000 + 800 + 90 + 4
>
> **Word form:**   thirty-five thousand eight hundred ninety-four

Write each number in standard form.

1.   70,000 + 1,000 + 600 + 90 + 2 = _71,692_

2.   800,000 + 60,000 + 900 + 30 + 7 = _860,937_

3.   twenty-four thousand eight hundred thirty-six = _24,836_

Write each number in expanded form.

4.   2,891 = _2,000 + 800 + 90 + 1_

5.   195,720 = _100,000 + 90,000 + 5,000 + 700 + 20_

6.   nine thousand four hundred seventy-five = _9,000 + 400 + 75_

7.   three hundred seventy thousand five hundred forty-one = _300,000 + 70,000 + 500 + 41_

Write each number in word form.

8.   991,347 = _Nine hundred nined-one thousand three hundred forty seven._

9.   30,724 = _Thirty thousand seven hundred twenty-four._

10.   20,000 + 4,000 + 700 + 30 + 1 = _Twenty-four thousand seven hundred thirty-one._

Exploring Place Value

Write each number in standard form, expanded form, or word form.

1. 40 + 3,000 + 1 + 800 + 70,000 in standard form is _____

2. 60,000 + 2,000 + 50 + 2 in word form is _____

3. 904,023 in expanded form is _____

4. 874,450 in expanded form is _____

5. five hundred two thousand eleven in standard form is _____

6. 100,034 in word form is _____

7. 600 + 700,000 + 9,000 in word form is _____

    _____

8. twelve thousand twelve in expanded form is _____

9. six hundred forty thousand two hundred nine in standard form is _____

10. one million in standard form is _____

11. 801,060 in expanded form is _____

12. twenty-six thousand five hundred ten in expanded form is _____

13. 1 + 10,000 + 10 + 100,000 in word form is _____

    _____

14. 202,202 in word form is _____

15. 671,000 in expanded form is _____

16. 500,000 + 7 in word form is _____

17. thirty-three thousand three in standard form is _____

Comparing Numbers

To compare numbers is to decide if the first number is **greater than** (>) or **less than** (<) the second number.

Compare numbers by place value, starting on the left. Moving from left to right, compare digits in each place in order. Stop at the first place where the numbers are different.

Compare 346 and 352.

3̲46   the digits in the first place   34̲6   the digits in the second place
       are the same: 3 hundreds   are different: 4 tens and 5 tens

3̲52   35̲2

4 tens < 5 tens, so 346 < 352

If the numbers are the same in all places, the numbers are **equal** (=).

Use >, <, or = to compare the numbers.

 1.   48 (<) 68

 2.   97,364 (<) 97,346

 3.   522 (>) 323

 4.   2,560 (>) 2,506

 5.   1,785 (>) 1,158

 6.   12,893 (=) 12,893

 7.   6,892 (=) 6,892

 8.   304,610 (<) 340,610

 9.   34,896 (<) 43,876

 10. 711,118 (<) 711,181

Comparing Numbers

Remember: Compare numbers by looking at each place value from left to right.

Use >, < , or = to compare the numbers.

1. 51 $>$ 31

2. 419 $>$ 411

3. 602 $>$ 206

4. 35,267 $=$ 35,267

5. 2,470 $>$ 2,047

6. 72,380 $<$ 72,387

7. 760,355 $>$ 50,366

8. 21,360 $>$ 21,306

9. 6,642 $<$ 66,403

10. 631,207 $>$ 62,746

11. 300,007 $<$ 300,008

12. 9,731 $>$ 973

13. 4,388 $=$ 4,388

14. 7,499 $<$ 7,500

15. 85,104 $<$ 95,104

16. 1,877 $>$ 1,766

17. 1,347 $<$ 1,374

18. 204,963 $>$ 201,652

Comparing Numbers

Use >, < , or = to compare the numbers.

1. 28,941 ◯ 28,914   2. 38,205 ◯ 3,205   3. 620,488 ◯ 628,048

4. 130,003 ◯ 130,030   5. 900,267 ◯ 900,267   6. 100,480 ◯ 1,000,000

7. 57,352 ◯ 75,352   8. 741,970 ◯ 741,790   9. 412,214 ◯ 412,412

10. 891,745 ◯ 891,145   11. 16,661 ◯ 106,661   12. 88,083 ◯ 83,083

13. 40,000 + 2,000 + 700 + 1 ◯ 40,000 + 2,000 + 700 + 1

14. 600,000 + 80,000 + 400 + 30 ◯ 60,000 + 800,000 + 400 + 30

15. 7,000 + 500,000 + 100 + 20 + 9 ◯ 20 + 7,000 + 100,000 + 500 + 9

16. 300,000 + 10,000 + 6,000 + 90 ◯ 90 + 6,000 + 100 + 300,000

17. 2 + 50 + 800 + 7,000 + 90,000 + 400,000 ◯ 400,000 + 800 + 50 + 90,000 + 2 + 7,000

18. three hundred thousand forty-five ◯ three hundred forty-five thousand

19. twenty-eight thousand six hundred one ◯ twenty-eight thousand one hundred six

20. nine hundred thirteen thousand three hundred nineteen ◯ nine hundred thirty thousand three hundred nine

Rounding Numbers

> **Rounding numbers** is a way of replacing one number with another number that tells about how many or how much.
>
> When rounding to a place value, look at the place value right before it. If it is 0, 1, 2, 3, or 4, round down. If it is 5, 6, 7, 8, or 9, round up.
>
> | Round 23 to the nearest ten. | Round 284 to the nearest hundred. |
> | :---: | :---: |
> | Look at the ones digit. | Look at the tens digit. |
> | 20   2<u>3</u>   30 | 200   2<u>8</u>4   300 |
> | Round 23 down to 20. | Round 284 up to 300. |

Round the numbers in each set to the place value listed.

1. ten

    24 _____   567 _____   7,075 _____

2. hundred

    26,483 _____   809 _____   4,095 _____

3. thousand

    873,609 _____   55,937 _____   2,148 _____

4. ten thousand

    902,813 _____   66,408 _____   40,742 _____

5. hundred thousand

    720,311 _____   485,407 _____   673,054 _____

Rounding Numbers

> Remember: If the digit to the right is less than 5, round down.
> If the digit to the right is 5 or greater, round up.

Round the numbers to the underlined place value.

1. 5<u>8</u>2 _____

2. <u>6</u>5,914 _____

3. 5<u>7</u>2,133 _____

4. 3<u>1</u>9 _____

5. 87,<u>4</u>08 _____

6. 60,8<u>4</u>7 _____

7. <u>7</u>55 _____

8. 3<u>2</u>5,035 _____

9. <u>4</u>92,795 _____

10. 9,3<u>2</u>1 _____

11. 98<u>2</u>,485 _____

12. 381,<u>2</u>91 _____

13. 8,<u>0</u>92 _____

14. <u>6</u>45,062 _____

15. <u>5</u>9,999 _____

16. <u>4</u>,902 _____

17. 768,<u>8</u>91 _____

18. 10<u>9</u>,900 _____

19. 82,<u>7</u>30 _____

20. 20<u>4</u>,519 _____

21. <u>9</u>78,010 _____

Rounding Numbers

Round the numbers to each place value.

		hundred	thousand	hundred thousand
1.	128,549			
2.	395,305			
3.	802,381			
4.	654,728			
5.	930,960			
6.	419,283			
7.	728,675			
8.	592,014			
9.	265,529			
10.	139,993			
11.	539,102			
12.	984,241			
13.	426,009			
14.	737,092			
15.	643,598			
16.	849,251			
17.	489,543			
18.	265,877			
19.	100,035			
20.	574,892			

Adding and Subtracting Large Numbers

Sometimes it is necessary to regroup when adding or subtracting large numbers.

$$\begin{array}{r} 1 \\ 53 \\ + 19 \\ \hline 72 \end{array}$$

Regroup the number 12 into 1 ten and 2 ones.

Carry the 1 ten to the tens column. Finish by adding the tens.

$$\begin{array}{r} 4\ 13 \\ \cancel{53} \\ - 16 \\ \hline 37 \end{array}$$

6 is too big to subtract from 3. Borrow a ten and regroup it into 10 ones. Subtract the ones column. Subtract the tens column.

Solve each problem. Regroup when necessary.

1.	6,376 + 2,019	2.	2,393 + 4,392	3.	8,293 + 4,239	4.	3,768 + 5,949	5.	1,665 + 3,773	6.	2,343 + 7,328

7.	7,320 + 5,394	8.	9,347 + 7,323	9.	8,659 + 9,347	10.	3,424 + 9,483	11.	6,784 + 1,296	12.	4,392 + 4,959

13.	9,534 − 2,389	14.	5,464 − 2,756	15.	3,526 − 1,653	16.	3,354 − 2,328	17.	5,247 − 3,836	18.	8,456 − 3,462

19.	4,755 − 3,875	20.	7,243 − 2,376	21.	6,845 − 4,764	22.	5,935 − 3,837	23.	4,376 − 2,438	24.	9,122 − 4,547

Adding and Subtracting Large Numbers

When subtracting, you may need to borrow from a zero.
To do this, borrow to make the zero a ten. Then, borrow from the ten.

$$\begin{array}{r} 9 \\ 4\,10\,12 \\ \cancel{5}\cancel{0}\cancel{2} \\ -\,346 \\ \hline 156 \end{array}$$

Borrow from the 5 hundred to make 10 tens.

Then, borrow a ten to make 12 ones.

Solve each problem. Regroup when necessary.

1. 84,936 + 25,432	2. 79,675 + 14,283	3. 35,349 + 36,393	4. 26,434 + 16,398	5. 49,231 + 15,332	6. 37,221 + 22,418
7. 76,376 + 52,019	8. 82,393 + 74,392	9. 58,293 + 34,239	10. 43,768 + 15,949	11. 91,665 + 13,773	12. 22,343 + 27,328
13. 57,320 + 65,394	14. 49,347 + 77,323	15. 28,659 + 19,347	16. 43,768 + 15,949	17. 56,784 + 61,296	18. 74,392 + 44,959
19. 89,534 − 12,389	20. 75,464 − 22,756	21. 63,526 − 51,653	22. 93,354 − 42,328	23. 45,247 − 33,836	24. 28,456 − 13,462
25. 54,755 − 23,875	26. 77,243 − 52,376	27. 16,845 − 14,764	28. 65,935 − 23,837	29. 84,376 − 12,438	30. 89,122 − 64,547
31. 32,643 − 11,439	32. 53,765 − 23,498	33. 67,236 − 12,276	34. 87,340 − 55,364	35. 96,849 − 74,114	36. 67,414 − 42,838

Adding and Subtracting Large Numbers

Fill in the missing numbers to complete each problem.

1.
```
    4  6,  3  7 ☐
+  ☐  4,  6  7  4
─────────────────
    6  1, ☐  4  6
```

2.
```
    1 ☐, 0  8  9
+     3  7,  4 ☐  5
─────────────────
    5  0,  5  6 ☐
```

3.
```
   ☐ 6,  1  7  8
+     6  4,  4 ☐  8
─────────────────
    9  0, ☐  6  6
```

4.
```
    3  5,  4  3 ☐
+  1 ☐, 7  8  0
─────────────────
    5  2,  2 ☐  9
```

5.
```
    9 ☐, 3 ☐  7
+     1  9,  8  0  7
─────────────────
    1 ☐, 1  5  4
```

6.
```
    3  1  2,  8  5 ☐
+  3 ☐ 3,  5  8  7
─────────────────
   ☐ 8  6,  4 ☐  3
```

7.
```
    9  2,  8  5 ☐
-  ☐ 5,  2  7  9
─────────────────
    5  7, ☐  7  2
```

8.
```
   ☐ 3,  1  2  8
-     6  4, ☐  5  6
─────────────────
       8,  7  7 ☐
```

9.
```
    1  2 ☐, 5 ☐  0
-     5  6,  2  0 ☐
─────────────────
    7  2,  3  3  9
```

10.
```
    3  9  4,  2  7  1
- ☐ 0  2,  0  5  6
─────────────────
    9 ☐, 2  1 ☐
```

11.
```
    7  2, ☐  3  3
-     5  4,  8 ☐  5
─────────────────
   ☐ 7,  3  0  8
```

12.
```
    8 ☐ 9,  0  2  2
-  4  6  7,  1  3 ☐
─────────────────
    3  8 ☐, 8  8  3
```

Multiplying Multi-Digit Numbers by One-Digit Numbers

Sometimes it is necessary to regroup when multiplying multi-digit numbers by one-digit numbers. Regroup by carrying tens.

$$\begin{array}{r} 2 \\ 28 \\ \times\ 3 \\ \hline 84 \end{array}$$

$8 \times 3 = 24$

Regroup the number 24 into 2 tens and 4 ones.

Carry the 2 tens to the tens column.

Multiply the tens column. $(2 \times 3 = 6)$

Then, add the carried tens. $(6 + 2 = 8)$

Solve each problem. Regroup when necessary.

1. $\begin{array}{r} 45 \\ \times\ 2 \\ \hline \end{array}$
2. $\begin{array}{r} 56 \\ \times\ 4 \\ \hline \end{array}$
3. $\begin{array}{r} 34 \\ \times\ 3 \\ \hline \end{array}$
4. $\begin{array}{r} 57 \\ \times\ 4 \\ \hline \end{array}$
5. $\begin{array}{r} 28 \\ \times\ 3 \\ \hline \end{array}$
6. $\begin{array}{r} 46 \\ \times\ 6 \\ \hline \end{array}$

7. $\begin{array}{r} 39 \\ \times\ 6 \\ \hline \end{array}$
8. $\begin{array}{r} 19 \\ \times\ 8 \\ \hline \end{array}$
9. $\begin{array}{r} 36 \\ \times\ 6 \\ \hline \end{array}$
10. $\begin{array}{r} 76 \\ \times\ 5 \\ \hline \end{array}$
11. $\begin{array}{r} 44 \\ \times\ 5 \\ \hline \end{array}$
12. $\begin{array}{r} 75 \\ \times\ 2 \\ \hline \end{array}$

13. $\begin{array}{r} 27 \\ \times\ 6 \\ \hline \end{array}$
14. $\begin{array}{r} 22 \\ \times\ 9 \\ \hline \end{array}$
15. $\begin{array}{r} 83 \\ \times\ 6 \\ \hline \end{array}$
16. $\begin{array}{r} 87 \\ \times\ 2 \\ \hline \end{array}$
17. $\begin{array}{r} 37 \\ \times\ 3 \\ \hline \end{array}$
18. $\begin{array}{r} 49 \\ \times\ 7 \\ \hline \end{array}$

19. $\begin{array}{r} 74 \\ \times\ 3 \\ \hline \end{array}$
20. $\begin{array}{r} 37 \\ \times\ 6 \\ \hline \end{array}$
21. $\begin{array}{r} 53 \\ \times\ 5 \\ \hline \end{array}$
22. $\begin{array}{r} 68 \\ \times\ 5 \\ \hline \end{array}$
23. $\begin{array}{r} 38 \\ \times\ 4 \\ \hline \end{array}$
24. $\begin{array}{r} 77 \\ \times\ 8 \\ \hline \end{array}$

Multiplying Multi-Digit Numbers by One-Digit Numbers

Solve each problem. Regroup when necessary.

1. $\begin{array}{r} 323 \\ \times\ \ 5 \\ \hline \end{array}$
2. $\begin{array}{r} 515 \\ \times\ \ 4 \\ \hline \end{array}$
3. $\begin{array}{r} 255 \\ \times\ \ 4 \\ \hline \end{array}$
4. $\begin{array}{r} 915 \\ \times\ \ 2 \\ \hline \end{array}$
5. $\begin{array}{r} 860 \\ \times\ \ 2 \\ \hline \end{array}$
6. $\begin{array}{r} 561 \\ \times\ \ 9 \\ \hline \end{array}$

7. $\begin{array}{r} 109 \\ \times\ \ 4 \\ \hline \end{array}$
8. $\begin{array}{r} 812 \\ \times\ \ 8 \\ \hline \end{array}$
9. $\begin{array}{r} 503 \\ \times\ \ 3 \\ \hline \end{array}$
10. $\begin{array}{r} 827 \\ \times\ \ 3 \\ \hline \end{array}$
11. $\begin{array}{r} 122 \\ \times\ \ 8 \\ \hline \end{array}$
12. $\begin{array}{r} 523 \\ \times\ \ 6 \\ \hline \end{array}$

13. $\begin{array}{r} 206 \\ \times\ \ 5 \\ \hline \end{array}$
14. $\begin{array}{r} 617 \\ \times\ \ 7 \\ \hline \end{array}$
15. $\begin{array}{r} 134 \\ \times\ \ 6 \\ \hline \end{array}$
16. $\begin{array}{r} 4,364 \\ \times\ \ \ \ 2 \\ \hline \end{array}$
17. $\begin{array}{r} 8,436 \\ \times\ \ \ \ 5 \\ \hline \end{array}$
18. $\begin{array}{r} 5,691 \\ \times\ \ \ \ 5 \\ \hline \end{array}$

19. $\begin{array}{r} 1,029 \\ \times\ \ \ \ 5 \\ \hline \end{array}$
20. $\begin{array}{r} 5,414 \\ \times\ \ \ \ 2 \\ \hline \end{array}$
21. $\begin{array}{r} 6,501 \\ \times\ \ \ \ 7 \\ \hline \end{array}$
22. $\begin{array}{r} 2,897 \\ \times\ \ \ \ 4 \\ \hline \end{array}$
23. $\begin{array}{r} 7,152 \\ \times\ \ \ \ 4 \\ \hline \end{array}$
24. $\begin{array}{r} 4,646 \\ \times\ \ \ \ 9 \\ \hline \end{array}$

25. $\begin{array}{r} 5,678 \\ \times\ \ \ \ 2 \\ \hline \end{array}$
26. $\begin{array}{r} 4,610 \\ \times\ \ \ \ 5 \\ \hline \end{array}$
27. $\begin{array}{r} 5,129 \\ \times\ \ \ \ 5 \\ \hline \end{array}$
28. $\begin{array}{r} 3,162 \\ \times\ \ \ \ 4 \\ \hline \end{array}$
29. $\begin{array}{r} 7,109 \\ \times\ \ \ \ 6 \\ \hline \end{array}$
30. $\begin{array}{r} 4,862 \\ \times\ \ \ \ 7 \\ \hline \end{array}$

Multiplying Multi-Digit Numbers by One-Digit Numbers

Solve each problem. Regroup when necessary.

1.   4,113
   ×   6

2.   7,312
   ×   7
   51,184

3.   8,900
   ×   8

4.   5,308
   ×   4
   21,232

5.   4,930
   ×   4

6.   6,342
   ×   5

7.   4,213
   ×   6

8.   9,980
   ×   5

9.   2,794
   ×   7

10.   9,755
   ×   8

11.   3,214
   ×   7

12.   2,317
   ×   3

13.   6,746
   ×   3

14.   6,677
   ×   4

15.   8,227
   ×   2

16.   5,857
   ×   3

17.   3,351
   ×   5

18.   2,356
   ×   3

19.   4,845
   ×   2

20.   7,934
   ×   3

21.   4,065
   ×   6

22.   2,132
   ×   6

23.   7,021
   ×   4

24.   9,442
   ×   3
   28,326

25.   4,365
   ×   6
   26,190

26.   3,225
   ×   5

27.   8,222
   ×   4
   32,888

28.   7,422
   ×   5
   37,110

29.   8,265
   ×   3
   24,795

30.   7,120
   ×   2

31.   7,322
   ×   6

32.   6,434
   ×   6
   38,604

33.   7,387
   ×   5

34.   8,483
   ×   4
   33,932

35.   8,612
   ×   6
   51,672

36.   6,947
   ×   5

Multiplying Two-Digit Numbers by Two-Digit Numbers

To multiply a two-digit number by another two-digit number, first multiply each digit by the ones column.

Next, put a 0 in the ones column under the 6. Then, multiply each digit by the tens column.

Last, add the two products together.

$$
\begin{array}{r} 23 \\ \times\ 12 \\ \hline 6 \end{array}
\longrightarrow
\begin{array}{r} 23 \\ \times\ 12 \\ \hline 46 \end{array}
$$

$$
\begin{array}{r} 23 \\ \times\ 12 \\ \hline 46 \\ 0 \end{array}
\qquad
\begin{array}{r} 23 \\ \times\ 12 \\ \hline 46 \\ 30 \end{array}
\qquad
\begin{array}{r} 23 \\ \times\ 12 \\ \hline 46 \\ +\ 230 \\ \hline 276 \end{array}
$$

Sometimes it is necessary to regroup.

Multiply by the ones digit. Regroup as needed.	Multiply by the tens digit. Regroup as needed.	Add the products together.
$\begin{array}{r} ^2\ \\ 37 \\ \times\ 24 \\ \hline 148 \end{array}$	$\begin{array}{r} ^1\ \\ 37 \\ \times\ 24 \\ \hline 148 \\ 740 \end{array}$	$\begin{array}{r} 37 \\ \times\ 24 \\ \hline 148 \\ +\ 740 \\ \hline 888 \end{array}$

Solve each problem. Regroup when necessary.

1.
$$\begin{array}{r} 58 \\ \times\ 26 \\ \hline \end{array}$$

2.
$$\begin{array}{r} 74 \\ \times\ 49 \\ \hline \end{array}$$

3.
$$\begin{array}{r} 69 \\ \times\ 27 \\ \hline \end{array}$$

4.
$$\begin{array}{r} 57 \\ \times\ 44 \\ \hline \end{array}$$

5.
$$\begin{array}{r} 44 \\ \times\ 37 \\ \hline \end{array}$$

6.
$$\begin{array}{r} 28 \\ \times\ 37 \\ \hline \end{array}$$

7.
$$\begin{array}{r} 45 \\ \times\ 36 \\ \hline \end{array}$$

8.
$$\begin{array}{r} 32 \\ \times\ 49 \\ \hline \end{array}$$

9.
$$\begin{array}{r} 59 \\ \times\ 30 \\ \hline \end{array}$$

10.
$$\begin{array}{r} 67 \\ \times\ 85 \\ \hline \end{array}$$

11.
$$\begin{array}{r} 52 \\ \times\ 47 \\ \hline \end{array}$$

12.
$$\begin{array}{r} 54 \\ \times\ 72 \\ \hline \end{array}$$

Multiplying Two-Digit Numbers by Two-Digit Numbers

Solve each problem. Regroup when necessary.

1. $\begin{array}{r} 41 \\ \times\ 18 \\ \hline \end{array}$
2. $\begin{array}{r} 53 \\ \times\ 38 \\ \hline \end{array}$
3. $\begin{array}{r} 73 \\ \times\ 46 \\ \hline \end{array}$
4. $\begin{array}{r} 42 \\ \times\ 30 \\ \hline \end{array}$
5. $\begin{array}{r} 86 \\ \times\ 75 \\ \hline \end{array}$

6. $\begin{array}{r} 38 \\ \times\ 22 \\ \hline \end{array}$
7. $\begin{array}{r} 36 \\ \times\ 12 \\ \hline \end{array}$
8. $\begin{array}{r} 62 \\ \times\ 44 \\ \hline \end{array}$
9. $\begin{array}{r} 81 \\ \times\ 72 \\ \hline \end{array}$
10. $\begin{array}{r} 56 \\ \times\ 13 \\ \hline \end{array}$

11. $\begin{array}{r} 64 \\ \times\ 47 \\ \hline \end{array}$
12. $\begin{array}{r} 82 \\ \times\ 51 \\ \hline \end{array}$
13. $\begin{array}{r} 25 \\ \times\ 17 \\ \hline \end{array}$
14. $\begin{array}{r} 91 \\ \times\ 43 \\ \hline \end{array}$
15. $\begin{array}{r} 49 \\ \times\ 28 \\ \hline \end{array}$

16. $\begin{array}{r} 68 \\ \times\ 32 \\ \hline \end{array}$
17. $\begin{array}{r} 42 \\ \times\ 18 \\ \hline \end{array}$
18. $\begin{array}{r} 86 \\ \times\ 42 \\ \hline \end{array}$
19. $\begin{array}{r} 35 \\ \times\ 28 \\ \hline \end{array}$
20. $\begin{array}{r} 73 \\ \times\ 56 \\ \hline \end{array}$

21. $\begin{array}{r} 72 \\ \times\ 43 \\ \hline \end{array}$
22. $\begin{array}{r} 58 \\ \times\ 63 \\ \hline \end{array}$
23. $\begin{array}{r} 83 \\ \times\ 27 \\ \hline \end{array}$
24. $\begin{array}{r} 70 \\ \times\ 60 \\ \hline \end{array}$
25. $\begin{array}{r} 54 \\ \times\ 27 \\ \hline \end{array}$

Multiplying Two-Digit Numbers by Two-Digit Numbers

Solve each problem. Regroup when necessary.

1. $\begin{array}{r} 48 \\ \times\ 38 \\ \hline \end{array}$
2. $\begin{array}{r} 63 \\ \times\ 73 \\ \hline \end{array}$
3. $\begin{array}{r} 67 \\ \times\ 24 \\ \hline \end{array}$
4. $\begin{array}{r} 89 \\ \times\ 24 \\ \hline \end{array}$
5. $\begin{array}{r} 55 \\ \times\ 63 \\ \hline \end{array}$
6. $\begin{array}{r} 39 \\ \times\ 28 \\ \hline \end{array}$

7. $\begin{array}{r} 51 \\ \times\ 40 \\ \hline \end{array}$
8. $\begin{array}{r} 48 \\ \times\ 69 \\ \hline \end{array}$
9. $\begin{array}{r} 58 \\ \times\ 73 \\ \hline \end{array}$
10. $\begin{array}{r} 73 \\ \times\ 28 \\ \hline \end{array}$
11. $\begin{array}{r} 55 \\ \times\ 33 \\ \hline \end{array}$
12. $\begin{array}{r} 88 \\ \times\ 62 \\ \hline \end{array}$

13. $\begin{array}{r} 34 \\ \times\ 66 \\ \hline \end{array}$
14. $\begin{array}{r} 62 \\ \times\ 44 \\ \hline \end{array}$
15. $\begin{array}{r} 68 \\ \times\ 59 \\ \hline \end{array}$
16. $\begin{array}{r} 27 \\ \times\ 45 \\ \hline \end{array}$
17. $\begin{array}{r} 29 \\ \times\ 89 \\ \hline \end{array}$
18. $\begin{array}{r} 53 \\ \times\ 24 \\ \hline \end{array}$

19. $\begin{array}{r} 28 \\ \times\ 48 \\ \hline \end{array}$
20. $\begin{array}{r} 70 \\ \times\ 47 \\ \hline \end{array}$
21. $\begin{array}{r} 50 \\ \times\ 42 \\ \hline \end{array}$
22. $\begin{array}{r} 38 \\ \times\ 22 \\ \hline \end{array}$
23. $\begin{array}{r} 45 \\ \times\ 56 \\ \hline \end{array}$
24. $\begin{array}{r} 62 \\ \times\ 46 \\ \hline \end{array}$

25. $\begin{array}{r} 76 \\ \times\ 49 \\ \hline \end{array}$
26. $\begin{array}{r} 66 \\ \times\ 38 \\ \hline \end{array}$
27. $\begin{array}{r} 37 \\ \times\ 48 \\ \hline \end{array}$
28. $\begin{array}{r} 67 \\ \times\ 49 \\ \hline \end{array}$
29. $\begin{array}{r} 67 \\ \times\ 81 \\ \hline \end{array}$
30. $\begin{array}{r} 47 \\ \times\ 86 \\ \hline \end{array}$

31. $\begin{array}{r} 48 \\ \times\ 29 \\ \hline \end{array}$
32. $\begin{array}{r} 45 \\ \times\ 28 \\ \hline \end{array}$
33. $\begin{array}{r} 32 \\ \times\ 62 \\ \hline \end{array}$
34. $\begin{array}{r} 58 \\ \times\ 26 \\ \hline \end{array}$
35. $\begin{array}{r} 74 \\ \times\ 49 \\ \hline \end{array}$
36. $\begin{array}{r} 69 \\ \times\ 27 \\ \hline \end{array}$

Dividing without Remainders

To divide multi-digit dividends by a one-digit divisor, see if the first digit is large enough to divide into. If yes, divide. Then, multiply the partial quotient by the divisor and subtract.

Next, bring down the second digit.

Then, divide the divisor into that number, multiply, and subtract.

Continue to bring down the next digit, divide the divisor into that number, multiply, and subtract.

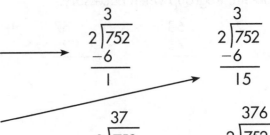

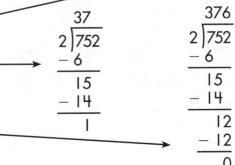

Solve each problem.

1. $4\overline{)72}$ 2. $3\overline{)81}$ 3. $7\overline{)98}$ 4. $5\overline{)90}$ 5. $6\overline{)84}$

6. $6\overline{)978}$ 7. $7\overline{)938}$ 8. $3\overline{)591}$ 9. $5\overline{)695}$ 10. $4\overline{)676}$

11. $4\overline{)712}$ 12. $5\overline{)880}$ 13. $2\overline{)918}$ 14. $3\overline{)891}$ 15. $3\overline{)792}$

Dividing without Remainders

Sometimes the first digit in the dividend is not large enough to divide into. Move to the next digit in the dividend. Divide into that two-digit dividend, multiply, and subtract.

$$
\begin{array}{r}
34 \\
6\overline{)204} \\
-18 \\
\hline
24 \\
-24 \\
\hline
0
\end{array}
$$

Solve each problem.

1. $5\overline{)415}$ 2. $3\overline{)156}$ 3. $7\overline{)154}$ 4. $8\overline{)696}$

5. $4\overline{)612}$ 6. $8\overline{)784}$ 7. $4\overline{)324}$ 8. $4\overline{)696}$

9. $6\overline{)426}$ 10. $3\overline{)159}$ 11. $4\overline{)356}$ 12. $5\overline{)475}$

13. $2\overline{)132}$ 14. $6\overline{)354}$ 15. $6\overline{)372}$ 16. $2\overline{)148}$

17. $4\overline{)348}$ 18. $2\overline{)146}$ 19. $5\overline{)315}$ 20. $4\overline{)320}$

Dividing without Remainders

Solve each problem.

1. $9\overline{)1,368}$ 2. $4\overline{)1,228}$ 3. $8\overline{)5,392}$ 4. $6\overline{)1,878}$

5. $5\overline{)1,395}$ 6. $7\overline{)2,926}$ 7. $4\overline{)1,008}$ 8. $5\overline{)975}$

9. $4\overline{)2,128}$ 10. $2\overline{)1,224}$ 11. $6\overline{)2,706}$ 12. $3\overline{)2,019}$

13. $3\overline{)1,008}$ 14. $8\overline{)3,888}$ 15. $7\overline{)1,421}$ 16. $5\overline{)1,125}$

17. $2\overline{)1,024}$ 18. $3\overline{)1,134}$ 19. $8\overline{)4,960}$ 20. $9\overline{)2,790}$

Dividing with Remainders

Sometimes it is necessary to name a remainder when dividing.
The **remainder** is the number remaining after the division is complete.

3 cannot divide into 2. No more digits can be brought down
from the dividend. The difference becomes the remainder (r).

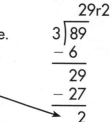

$$\begin{array}{r} 29r2 \\ 3\overline{)89} \\ -6 \\ \hline 29 \\ -27 \\ \hline 2 \end{array}$$

Solve each problem.

1. $7\overline{)82}$

2. $4\overline{)54}$

3. $3\overline{)26}$

4. $8\overline{)95}$

5. $4\overline{)18}$

6. $7\overline{)57}$

7. $4\overline{)63}$

8. $5\overline{)22}$

9. $5\overline{)18}$

10. $5\overline{)81}$

11. $4\overline{)41}$

12. $3\overline{)29}$

13. $6\overline{)74}$

14. $8\overline{)37}$

15. $5\overline{)42}$

Dividing with Remainders

Solve each problem.

1. $4\overline{)873}$　　　　2. $5\overline{)943}$　　　　3. $8\overline{)957}$　　　　4. $9\overline{)987}$

5. $7\overline{)915}$　　　　6. $5\overline{)527}$　　　　7. $2\overline{)597}$　　　　8. $9\overline{)973}$

9. $4\overline{)574}$　　　　10. $6\overline{)653}$　　　　11. $3\overline{)784}$　　　　12. $4\overline{)486}$

13. $3\overline{)629}$　　　　14. $2\overline{)301}$　　　　15. $5\overline{)637}$　　　　16. $4\overline{)862}$

17. $2\overline{)733}$　　　　18. $8\overline{)937}$　　　　19. $3\overline{)574}$　　　　20. $4\overline{)653}$

Dividing with Remainders

Solve each problem. Show your work on another sheet of paper. Write your answers here.

1. $6\overline{)7,391}$

2. $3\overline{)2,874}$

3. $3\overline{)6,238}$

4. $8\overline{)4,376}$

5. $6\overline{)3,764}$

6. $9\overline{)2,819}$

7. $2\overline{)8,497}$

8. $6\overline{)8,149}$

9. $5\overline{)3,381}$

10. $4\overline{)2,987}$

11. $7\overline{)8,040}$

12. $3\overline{)3,788}$

13. $7\overline{)5,001}$

14. $2\overline{)6,841}$

15. $6\overline{)9,469}$

16. $5\overline{)5,328}$

17. $5\overline{)7,384}$

18. $4\overline{)5,978}$

19. $4\overline{)1,538}$

20. $2\overline{)4,811}$

21. $7\overline{)8,598}$

22. $4\overline{)8,572}$

23. $3\overline{)6,943}$

24. $6\overline{)6,436}$

25. $8\overline{)4,687}$

26. $5\overline{)5,237}$

27. $7\overline{)4,790}$

28. $5\overline{)3,486}$

29. $4\overline{)9,035}$

30. $7\overline{)4,001}$

Finding Equivalent Fractions

$$\frac{1}{2} = \frac{2}{4}$$

Fractions that equal the same amount are called **equivalent fractions**.

$$\frac{1}{2} = \frac{2}{4}$$

It is the same amount. The pieces are just different sizes.

Write the equivalent fractions.

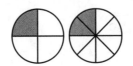

1. _____ = _____

2. _____ = _____

3. _____ = _____

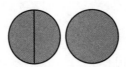

4. _____ = _____

5. _____ = _____

6. _____ = _____

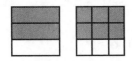

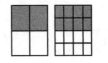

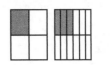

7. _____ = _____

8. _____ = _____

9. _____ = _____

10. _____ = _____

11. _____ = _____

12. _____ = _____

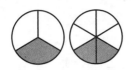

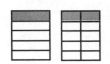

13. _____ = _____

14. _____ = _____

15. _____ = _____

Finding Equivalent Fractions

> To find equivalent fractions, multiply the numerator and the denominator by the same number.

Write the missing numerators to make the fractions in each row equivalent. Draw one pair of equivalent fractions for each problem.

1. $\dfrac{1}{2} = \dfrac{}{36} = \dfrac{}{18} = \dfrac{}{16} = \dfrac{}{42} = \dfrac{}{48}$

2. $\dfrac{5}{6} = \dfrac{}{48} = \dfrac{}{12} = \dfrac{}{30} = \dfrac{}{18} = \dfrac{}{24}$

3. $\dfrac{1}{3} = \dfrac{}{9} = \dfrac{}{27} = \dfrac{}{90} = \dfrac{}{6} = \dfrac{}{12}$

4. $\dfrac{3}{4} = \dfrac{}{24} = \dfrac{}{16} = \dfrac{}{8} = \dfrac{}{20} = \dfrac{}{36}$

5. $\dfrac{4}{9} = \dfrac{}{18} = \dfrac{}{45} = \dfrac{}{36} = \dfrac{}{54} = \dfrac{}{27}$

6. $\dfrac{7}{8} = \dfrac{}{16} = \dfrac{}{56} = \dfrac{}{24} = \dfrac{}{48} = \dfrac{}{32}$

7. $\dfrac{3}{7} = \dfrac{}{21} = \dfrac{}{42} = \dfrac{}{14} = \dfrac{}{35} = \dfrac{}{28}$

8. $\dfrac{2}{5} = \dfrac{}{50} = \dfrac{}{10} = \dfrac{}{40} = \dfrac{}{15} = \dfrac{}{25}$

Finding Equivalent Fractions

Write the missing numerator to make each pair equivalent.

1. $\dfrac{2}{3} = \dfrac{}{12}$

2. $\dfrac{8}{9} = \dfrac{}{54}$

3. $\dfrac{1}{2} = \dfrac{}{10}$

4. $\dfrac{1}{8} = \dfrac{}{32}$

5. $\dfrac{4}{9} = \dfrac{}{81}$

6. $\dfrac{2}{9} = \dfrac{}{18}$

7. $\dfrac{3}{4} = \dfrac{}{16}$

8. $\dfrac{1}{2} = \dfrac{}{12}$

9. $\dfrac{4}{5} = \dfrac{}{25}$

10. $\dfrac{2}{5} = \dfrac{}{30}$

11. $\dfrac{7}{8} = \dfrac{}{64}$

12. $\dfrac{2}{3} = \dfrac{}{15}$

13. $\dfrac{2}{5} = \dfrac{}{10}$

14. $\dfrac{3}{8} = \dfrac{}{16}$

15. $\dfrac{5}{8} = \dfrac{}{24}$

16. $\dfrac{3}{4} = \dfrac{}{24}$

17. $\dfrac{3}{5} = \dfrac{}{15}$

18. $\dfrac{3}{7} = \dfrac{}{14}$

19. $\dfrac{1}{6} = \dfrac{}{12}$

20. $\dfrac{4}{5} = \dfrac{}{20}$

21. $\dfrac{3}{7} = \dfrac{}{21}$

22. $\dfrac{5}{6} = \dfrac{}{42}$

23. $\dfrac{1}{6} = \dfrac{}{36}$

24. $\dfrac{5}{8} = \dfrac{}{40}$

Reducing Fractions

To put a fraction in **simplest form** means to rename or **reduce** the fraction without changing the amount.

Follow the steps to reduce a fraction.

$\frac{6}{9}$ 1.   Find the largest number that can be divided into the numerator and denominator.

$\div 3$ 2.   Both numbers can be divided by 3. $\frac{6}{9} \div \frac{3}{3} = \frac{2}{3}$

$\frac{2}{3}$ 3.   The simplest form of $\frac{6}{9}$ is $\frac{2}{3}$.

   The two fractions still represent the same amount.

Reduce each fraction to simplest form.

1. $\frac{4}{10} \div \div = \underline{\ \ }$

2. $\frac{8}{32} \div \div = \underline{\ \ }$

3. $\frac{10}{12} \div \div = \underline{\ \ }$

4. $\frac{18}{27} \div \div = \underline{\ \ }$

5. $\frac{5}{15} \div \div = \underline{\ \ }$

6. $\frac{4}{26} \div \div = \underline{\ \ }$

7. $\frac{16}{56} \div \div = \underline{\ \ }$

8. $\frac{20}{45} \div \div = \underline{\ \ }$

9. $\frac{18}{40} \div \div = \underline{\ \ }$

Reducing Fractions

Write each fraction in simplest form.

1. $\dfrac{6}{8}$ =

2. $\dfrac{3}{24}$ =

3. $\dfrac{20}{35}$ =

4. $\dfrac{15}{20}$ =

5. $\dfrac{10}{20}$ =

6. $\dfrac{6}{16}$ =

7. $\dfrac{5}{20}$ =

8. $\dfrac{4}{8}$ =

9. $\dfrac{4}{16}$ =

10. $\dfrac{6}{9}$ =

11. $\dfrac{4}{20}$ =

12. $\dfrac{3}{15}$ =

13. $\dfrac{3}{12}$ =

14. $\dfrac{5}{15}$ =

15. $\dfrac{8}{16}$ =

16. $\dfrac{7}{21}$ =

17. $\dfrac{5}{25}$ =

18. $\dfrac{15}{30}$ =

19. $\dfrac{2}{8}$ =

20. $\dfrac{14}{21}$ =

21. $\dfrac{12}{16}$ =

Reducing Fractions

Write each fraction in simplest form.

1. $\frac{4}{8}$ =

2. $\frac{7}{14}$ =

3. $\frac{20}{30}$ =

4. $\frac{10}{28}$ =

5. $\frac{14}{40}$ =

6. $\frac{6}{20}$ =

7. $\frac{4}{12}$ =

8. $\frac{2}{8}$ =

9. $\frac{5}{30}$ =

10. $\frac{3}{9}$ =

11. $\frac{2}{6}$ =

12. $\frac{3}{15}$ =

13. $\frac{3}{12}$ =

14. $\frac{8}{24}$ =

15. $\frac{8}{20}$ =

16. $\frac{6}{18}$ =

17. $\frac{5}{20}$ =

18. $\frac{15}{20}$ =

19. $\frac{2}{4}$ =

20. $\frac{15}{21}$ =

21. $\frac{12}{30}$ =

22. $\frac{20}{22}$ =

23. $\frac{7}{28}$ =

24. $\frac{16}{32}$ =

25. $\frac{12}{15}$ =

26. $\frac{18}{24}$ =

27. $\frac{4}{18}$ =

28. $\frac{5}{15}$ =

29. $\frac{15}{20}$ =

30. $\frac{21}{45}$ =

Writing Improper Fractions as Mixed Numbers

When the numerator is greater than or equal to the denominator, it is called an **improper fraction**.

$\frac{9}{4}$

When a whole number is with a fraction, it is called a **mixed number**.

$2\frac{1}{4}$

An improper fraction ($\frac{9}{4}$) can be changed to a mixed number.

Divide the numerator by the denominator.

The quotient becomes the whole number. The remainder becomes a fraction. Use the denominator of the improper fraction.

$$\frac{9}{4}$$

$$\begin{array}{r} 2r1 \\ 4\overline{)9} \\ -8 \\ \hline 1 \end{array}$$

$2\frac{1}{4}$

Write each improper fraction as a mixed number.

1. $\frac{14}{3}$ =

2. $\frac{22}{7}$ =

3. $\frac{44}{8}$ =

4. $\frac{32}{5}$ =

5. $\frac{13}{4}$ =

6. $\frac{40}{6}$ =

7. $\frac{18}{4}$ =

8. $\frac{59}{9}$ =

9. $\frac{15}{8}$ =

10. $\frac{23}{9}$ =

11. $\frac{32}{7}$ =

12. $\frac{44}{6}$ =

13. $\frac{12}{5}$ =

14. $\frac{13}{5}$ =

15. $\frac{58}{9}$ =

16. $\frac{15}{4}$ =

Writing Improper Fractions as Mixed Numbers

Write each improper fraction as a mixed number in simplest form.

1. $\dfrac{4}{3} =$

2. $\dfrac{20}{15} =$

3. $\dfrac{7}{4} =$

4. $\dfrac{55}{12} =$

5. $\dfrac{18}{5} =$

6. $\dfrac{5}{2} =$

7. $\dfrac{5}{3} =$

8. $\dfrac{12}{5} =$

9. $\dfrac{13}{4} =$

10. $\dfrac{15}{6} =$

11. $\dfrac{13}{2} =$

12. $\dfrac{17}{9} =$

13. $\dfrac{10}{4} =$

14. $\dfrac{19}{2} =$

15. $\dfrac{27}{5} =$

16. $\dfrac{15}{4} =$

17. $\dfrac{8}{3} =$

18. $\dfrac{15}{8} =$

19. $\dfrac{6}{4} =$

20. $\dfrac{43}{7} =$

21. $\dfrac{19}{11} =$

22. $\dfrac{20}{7} =$

23. $\dfrac{9}{4} =$

24. $\dfrac{17}{4} =$

Writing Improper Fractions as Mixed Numbers

Write each improper fraction as a mixed number in simplest form.

1. $\dfrac{6}{4}$ =

2. $\dfrac{21}{12}$ =

3. $\dfrac{9}{4}$ =

4. $\dfrac{25}{11}$ =

5. $\dfrac{19}{5}$ =

6. $\dfrac{3}{2}$ =

7. $\dfrac{7}{4}$ =

8. $\dfrac{13}{3}$ =

9. $\dfrac{14}{6}$ =

10. $\dfrac{16}{5}$ =

11. $\dfrac{13}{5}$ =

12. $\dfrac{14}{8}$ =

13. $\dfrac{11}{2}$ =

14. $\dfrac{17}{4}$ =

15. $\dfrac{19}{2}$ =

16. $\dfrac{25}{3}$ =

17. $\dfrac{8}{3}$ =

18. $\dfrac{11}{6}$ =

19. $\dfrac{10}{3}$ =

20. $\dfrac{33}{6}$ =

21. $\dfrac{14}{9}$ =

22. $\dfrac{20}{8}$ =

23. $\dfrac{7}{4}$ =

24. $\dfrac{13}{3}$ =

25. $\dfrac{12}{5}$ =

26. $\dfrac{18}{11}$ =

27. $\dfrac{9}{2}$ =

28. $\dfrac{15}{4}$ =

29. $\dfrac{10}{6}$ =

30. $\dfrac{10}{4}$ =

Writing Mixed Numbers as Improper Fractions

When the numerator is greater than or equal to the denominator, it is called an **improper fraction**. $\frac{9}{4}$

When a whole number is with a fraction, it is called a **mixed number**. $2\frac{1}{4}$

A mixed number ($2\frac{1}{4}$) can be changed to an improper fraction.

Multiply the whole number by the denominator.	Add the product to the numerator.	Write the sum over the original denominator.
$2\frac{1}{4}$ $2 \times 4 = 8$	$8 + 1 = 9$	$\frac{9}{4}$

Write each mixed number as an improper fraction.

1. $1\frac{4}{3}$ =

2. $3\frac{1}{7}$ =

3. $4\frac{4}{8}$ =

4. $6\frac{2}{5}$ =

5. $3\frac{3}{4}$ =

6. $7\frac{4}{6}$ =

7. $5\frac{1}{4}$ =

8. $6\frac{5}{9}$ =

9. $1\frac{7}{8}$ =

10. $2\frac{5}{9}$ =

11. $4\frac{4}{7}$ =

12. $7\frac{2}{6}$ =

13. $2\frac{2}{5}$ =

14. $2\frac{3}{5}$ =

15. $6\frac{4}{9}$ =

16. $3\frac{3}{4}$ =

Writing Mixed Numbers as Improper Fractions

Write each mixed number as an improper fraction.

1. $3\frac{1}{2}$ =

2. $5\frac{7}{8}$ =

3. $7\frac{4}{5}$ =

4. $1\frac{1}{10}$ =

5. $6\frac{5}{8}$ =

6. $5\frac{2}{3}$ =

7. $9\frac{1}{2}$ =

8. $4\frac{3}{8}$ =

9. $8\frac{2}{3}$ =

10. $2\frac{2}{3}$ =

11. $2\frac{4}{9}$ =

12. $4\frac{3}{4}$ =

13. $2\frac{3}{8}$ =

14. $4\frac{2}{4}$ =

15. $6\frac{5}{7}$ =

16. $10\frac{3}{5}$ =

17. $4\frac{5}{9}$ =

18. $7\frac{5}{6}$ =

Writing Mixed Numbers as Improper Fractions

Write each mixed number as an improper fraction.

1. $1\frac{2}{3}$ =

2. $10\frac{5}{6}$ =

3. $5\frac{4}{5}$ =

4. $2\frac{1}{12}$ =

5. $12\frac{4}{5}$ =

6. $1\frac{1}{5}$ =

7. $7\frac{1}{6}$ =

8. $9\frac{6}{8}$ =

9. $9\frac{2}{8}$ =

10. $20\frac{2}{8}$ =

11. $8\frac{3}{9}$ =

12. $3\frac{3}{8}$ =

13. $2\frac{3}{4}$ =

14. $1\frac{1}{4}$ =

15. $15\frac{3}{4}$ =

16. $7\frac{10}{16}$ =

17. $11\frac{3}{9}$ =

18. $12\frac{13}{15}$ =

19. $3\frac{2}{5}$ =

20. $6\frac{1}{3}$ =

21. $7\frac{2}{5}$ =

Comparing Fractions

Use the fraction bars to compare the fractions.

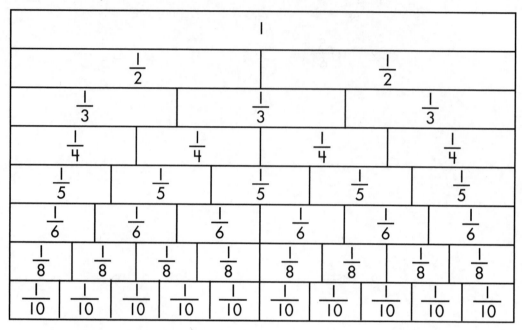

1. Circle the fraction that is less.

$\frac{1}{2}$	$\frac{2}{5}$
$\frac{2}{6}$	$\frac{2}{8}$
$\frac{4}{6}$	$\frac{3}{4}$
$\frac{4}{10}$	$\frac{1}{3}$

2. Circle the fraction that is greater.

$\frac{3}{6}$	$\frac{3}{5}$
$\frac{4}{8}$	$\frac{4}{5}$
$\frac{9}{10}$	$\frac{2}{3}$
$\frac{4}{10}$	$\frac{2}{6}$

Comparing Fractions

| To compare, you must find equivalent fractions.

$\dfrac{1}{4} \bigcirc \dfrac{1}{8}$ | First, we must find a common denominator. Does something multiplied by 4 equal 8? Yes, 2.

The common denominator is 8. | Multiply the numerator and denominator by 2 to create equivalent fractions.

$\dfrac{1 \times 2}{4 \times 2} = \dfrac{2}{8}$ | $\dfrac{1}{4} = \dfrac{2}{8}$

Compare.

$\dfrac{1}{4} > \dfrac{1}{8}$ |

Identify each fraction. Use > or < to compare the fractions.

1.

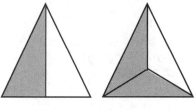

___ ◯ ___

2.

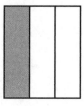

___ ◯ ___

3.

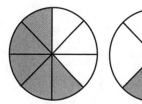

___ ◯ ___

4.

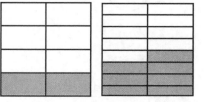

___ ◯ ___

5.

___ ◯ ___

6.

___ ◯ ___

7.

___ ◯ ___

8.

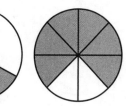

___ ◯ ___

Comparing Fractions

Use >, <, or = to compare the fractions.

1. $\dfrac{5}{10}$ ◯ $\dfrac{2}{10}$

2. $\dfrac{1}{3}$ ◯ $\dfrac{2}{3}$

3. $\dfrac{5}{8}$ ◯ $\dfrac{6}{8}$

4. $\dfrac{3}{10}$ ◯ $\dfrac{8}{10}$

5. $\dfrac{1}{4}$ ◯ $\dfrac{3}{4}$

6. $\dfrac{6}{7}$ ◯ $\dfrac{3}{7}$

7. $\dfrac{4}{6}$ ◯ $\dfrac{1}{6}$

8. $\dfrac{5}{9}$ ◯ $\dfrac{4}{9}$

9. $\dfrac{12}{24}$ ◯ $\dfrac{2}{4}$

10. $\dfrac{1}{2}$ ◯ $\dfrac{24}{50}$

11. $\dfrac{4}{5}$ ◯ $\dfrac{5}{6}$

12. $\dfrac{9}{12}$ ◯ $\dfrac{15}{24}$

13. $\dfrac{1}{2}$ ◯ $\dfrac{3}{4}$

14. $\dfrac{1}{6}$ ◯ $\dfrac{2}{3}$

15. $\dfrac{3}{4}$ ◯ $\dfrac{1}{8}$

16. $\dfrac{2}{4}$ ◯ $\dfrac{1}{2}$

17. $\dfrac{6}{8}$ ◯ $\dfrac{2}{8}$

18. $\dfrac{1}{3}$ ◯ $\dfrac{2}{9}$

19. $\dfrac{4}{6}$ ◯ $\dfrac{2}{3}$

20. $\dfrac{1}{5}$ ◯ $\dfrac{2}{15}$

21. $\dfrac{3}{4}$ ◯ $\dfrac{5}{8}$

Adding Fractions with Like Denominators

To add fractions with the same denominators, add the numerators.

Reduce the answer to simplest form.

$$\frac{3}{10} + \frac{3}{10} = \frac{6}{10}$$

$$\frac{6}{10} \div \frac{2}{2} = \frac{3}{5}$$

Solve each problem. Write the answer in simplest form.

1. $\frac{2}{5} + \frac{3}{5} =$

2. $\frac{2}{7} + \frac{4}{7} =$

3. $\frac{3}{8} + \frac{1}{8} =$

4. $\frac{2}{9} + \frac{3}{9} =$

5. $\frac{5}{8} + \frac{2}{8} =$

6. $\frac{3}{18} + \frac{1}{18} =$

7. $\frac{2}{5} + \frac{2}{5} =$

8. $\frac{1}{8} + \frac{1}{8} =$

9. $\frac{3}{4} + \frac{1}{4} =$

10. $\frac{11}{15} + \frac{10}{15} =$

11. $\frac{2}{25} + \frac{2}{25} =$

12. $\frac{3}{12} + \frac{1}{12} =$

Adding Fractions with Like Denominators

Solve each problem. Write the answer in simplest form.

1. $\frac{1}{3} + \frac{2}{3} =$

2. $\frac{2}{9} + \frac{5}{9} =$

3. $\frac{1}{6} + \frac{1}{6} =$

4. $\frac{3}{6} + \frac{1}{6} =$

5. $\frac{2}{4} + \frac{2}{4} =$

6. $\frac{1}{2} + \frac{1}{2} =$

7. $\frac{5}{8} + \frac{3}{8} =$

8. $\frac{1}{5} + \frac{2}{5} =$

9. $\frac{2}{10} + \frac{4}{10} =$

10. $\frac{1}{4} + \frac{1}{4} =$

11. $\frac{3}{5} + \frac{2}{5} =$

12. $\frac{3}{7} + \frac{2}{7} =$

13. $\frac{2}{3} + \frac{2}{3} =$

14. $\frac{1}{7} + \frac{1}{7} =$

15. $\frac{1}{6} + \frac{4}{6} =$

Adding Fractions with Like Denominators

Solve each problem. Write the answer in simplest form.

1. $\dfrac{2}{7}$
$+\dfrac{3}{7}$

2. $\dfrac{6}{8}$
$+\dfrac{1}{8}$

3. $\dfrac{7}{10}$
$+\dfrac{3}{10}$

4. $\dfrac{3}{7}$
$+\dfrac{1}{7}$

5. $\dfrac{1}{5}$
$+\dfrac{3}{5}$

6. $\dfrac{3}{5}$
$+\dfrac{2}{5}$

7. $\dfrac{1}{4}$
$+\dfrac{2}{4}$

8. $\dfrac{1}{5}$
$+\dfrac{3}{5}$

9. $\dfrac{4}{8}$
$+\dfrac{2}{8}$

10. $\dfrac{4}{7}$
$+\dfrac{5}{7}$

11. $\dfrac{1}{8}$
$+\dfrac{5}{8}$

12. $\dfrac{3}{8}$
$+\dfrac{4}{8}$

13. $\dfrac{2}{10}$
$+\dfrac{4}{10}$

14. $\dfrac{3}{4}$
$+\dfrac{1}{4}$

15. $\dfrac{1}{3}$
$+\dfrac{1}{3}$

16. $\dfrac{8}{9}$
$+\dfrac{8}{9}$

17. $\dfrac{2}{6}$
$+\dfrac{1}{6}$

18. $\dfrac{5}{12}$
$+\dfrac{5}{12}$

19. $\dfrac{1}{6}$
$+\dfrac{3}{6}$

20. $\dfrac{2}{9}$
$+\dfrac{1}{9}$

Subtracting Fractions with Like Denominators

To subtract fractions with the same denominators, subtract the numerators.

Reduce the answer to simplest form.

$$\frac{8}{10} - \frac{3}{10} = \frac{5}{10}$$

$$\frac{5}{10} \div \frac{5}{5} = \frac{1}{2}$$

Solve each problem. Write the answer in simplest form.

1. $\frac{2}{5} - \frac{1}{5} =$

2. $\frac{6}{7} - \frac{4}{7} =$

3. $\frac{3}{8} - \frac{1}{8} =$

4. $\frac{5}{9} - \frac{3}{9} =$

5. $\frac{5}{8} - \frac{2}{8} =$

6. $\frac{7}{8} - \frac{1}{8} =$

7. $\frac{4}{5} - \frac{2}{5} =$

8. $\frac{4}{8} - \frac{1}{8} =$

9. $\frac{3}{4} - \frac{1}{4} =$

10. $\frac{11}{15} - \frac{10}{15} =$

11. $\frac{3}{5} - \frac{2}{5} =$

12. $\frac{2}{4} - \frac{1}{4} =$

Subtracting Fractions with Like Denominators

Solve each problem. Write the answer in simplest form.

1. $\frac{2}{5} - \frac{1}{5} =$

2. $\frac{5}{9} - \frac{2}{9} =$

3. $\frac{6}{7} - \frac{1}{7} =$

4. $\frac{5}{6} - \frac{3}{6} =$

5. $\frac{2}{9} - \frac{2}{9} =$

6. $\frac{7}{9} - \frac{3}{9} =$

7. $\frac{5}{10} - \frac{2}{10} =$

8. $\frac{5}{5} - \frac{2}{5} =$

9. $\frac{9}{20} - \frac{2}{20} =$

10. $\frac{2}{2} - \frac{1}{2} =$

11. $\frac{3}{8} - \frac{2}{8} =$

12. $\frac{2}{3} - \frac{1}{3} =$

13. $\frac{3}{4} - \frac{2}{4} =$

14. $\frac{1}{7} - \frac{1}{7} =$

15. $\frac{9}{10} - \frac{7}{10} =$

Subtracting Fractions with Like Denominators

Solve each problem. Write the answer in simplest form.

1. $\dfrac{5}{6}$
    $-\dfrac{1}{6}$

2. $\dfrac{7}{12}$
    $-\dfrac{5}{12}$

3. $\dfrac{9}{14}$
    $-\dfrac{1}{14}$

4. $\dfrac{7}{8}$
    $-\dfrac{5}{8}$

5. $\dfrac{6}{8}$
    $-\dfrac{3}{8}$

6. $\dfrac{5}{7}$
    $-\dfrac{2}{7}$

7. $\dfrac{9}{11}$
    $-\dfrac{1}{11}$

8. $\dfrac{5}{9}$
    $-\dfrac{4}{9}$

9. $\dfrac{3}{10}$
    $-\dfrac{1}{10}$

10. $\dfrac{7}{9}$
    $-\dfrac{1}{9}$

11. $\dfrac{5}{8}$
    $-\dfrac{1}{8}$

12. $\dfrac{5}{7}$
    $-\dfrac{3}{7}$

13. $\dfrac{15}{16}$
    $-\dfrac{11}{16}$

14. $\dfrac{4}{5}$
    $-\dfrac{2}{5}$

15. $\dfrac{2}{3}$
    $-\dfrac{1}{3}$

16. $\dfrac{2}{5}$
    $-\dfrac{1}{5}$

17. $\dfrac{3}{4}$
    $-\dfrac{1}{4}$

18. $\dfrac{13}{15}$
    $-\dfrac{11}{15}$

19. $\dfrac{9}{10}$
    $-\dfrac{7}{10}$

20. $\dfrac{3}{3}$
    $-\dfrac{1}{3}$

Adding Mixed Numbers with Like Denominators

To add mixed numbers, add the whole numbers and fractions separately:

$$3\frac{1}{3} + 2\frac{1}{3} = (3 + 2) + (\frac{1}{3} + \frac{1}{3}) = 5\frac{2}{3}$$

OR

Write the mixed numbers as improper fractions and add. Then, change the answer back into a mixed number:

$$3\frac{1}{3} + 2\frac{1}{3} = \frac{10}{3} + \frac{7}{3} = \frac{17}{3} = 5\frac{2}{3}$$

Solve. Add the whole numbers and fractions separately.

1. $1\frac{1}{4} + 2\frac{2}{4} =$ (_____ + _____) + (— + —) = _____

2. $3\frac{5}{8} + 4\frac{2}{8} =$ (_____ + _____) + (— + —) = _____

3. $5\frac{2}{6} + 7\frac{3}{6} =$ (_____ + _____) + (— + —) = _____

4. $8\frac{1}{12} + 9\frac{5}{12} =$ (_____ + _____) + (— + —) = _____

Solve. Write the mixed numbers as improper fractions and add. Show your answer as a mixed number in simplest form.

5. $4\frac{3}{7} + 3\frac{2}{7} =$ _____ + _____ = _____ = _____

6. $6\frac{2}{8} + 5\frac{4}{8} =$ _____ + _____ = _____ = _____

7. $3\frac{4}{10} + 1\frac{1}{10} =$ _____ + _____ = _____ = _____

8. $4\frac{2}{5} + 2\frac{1}{5} =$ _____ + _____ = _____ = _____

Adding Mixed Numbers with Like Denominators

Remember: If the fractional parts add up to more than one whole, add the whole to the whole number part of the answer.

Solve each problem. Write the answer in simplest form.

1. $2\frac{2}{5} + 2\frac{3}{5} =$

2. $5\frac{2}{7} + 6\frac{4}{7} =$

3. $3\frac{3}{8} + 4\frac{1}{8} =$

4. $4\frac{2}{9} + 5\frac{3}{9} =$

5. $6\frac{5}{8} + 7\frac{2}{8} =$

6. $5\frac{3}{8} + 4\frac{2}{8} =$

7. $8\frac{2}{5} + 1\frac{2}{5} =$

8. $7\frac{1}{8} + 7\frac{1}{8} =$

9. $2\frac{3}{4} + 2\frac{1}{4} =$

10. $5\frac{11}{15} + 6\frac{4}{15} =$

11. $2\frac{2}{5} + 2\frac{2}{5} =$

12. $4\frac{3}{4} + 1\frac{1}{4} =$

Adding Mixed Numbers with Like Denominators

Solve each problem. Write the answer in simplest form.

1. $\begin{array}{r} 4\frac{5}{8} \\ + 5\frac{4}{8} \\ \hline \end{array}$

2. $\begin{array}{r} 2\frac{2}{5} \\ + 6\frac{4}{5} \\ \hline \end{array}$

3. $\begin{array}{r} 4\frac{5}{8} \\ + 5\frac{4}{8} \\ \hline \end{array}$

4. $\begin{array}{r} 10\frac{3}{4} \\ + 8\frac{2}{4} \\ \hline \end{array}$

5. $\begin{array}{r} 1\frac{1}{2} \\ + 4\frac{1}{2} \\ \hline \end{array}$

6. $\begin{array}{r} 8\frac{4}{9} \\ + 1\frac{5}{9} \\ \hline \end{array}$

7. $\begin{array}{r} 7\frac{6}{9} \\ + 2\frac{1}{9} \\ \hline \end{array}$

8. $\begin{array}{r} 2\frac{5}{6} \\ + 8\frac{5}{6} \\ \hline \end{array}$

9. $\begin{array}{r} 6\frac{2}{3} \\ + 7\frac{2}{3} \\ \hline \end{array}$

10. $\begin{array}{r} 4\frac{2}{7} \\ + 5\frac{3}{7} \\ \hline \end{array}$

11. $\begin{array}{r} 4\frac{2}{5} \\ + 6\frac{4}{5} \\ \hline \end{array}$

12. $\begin{array}{r} 9\frac{4}{12} \\ + 6\frac{10}{12} \\ \hline \end{array}$

13. $\begin{array}{r} 3\frac{9}{10} \\ + 7\frac{6}{10} \\ \hline \end{array}$

14. $\begin{array}{r} 3\frac{1}{3} \\ + 4\frac{2}{3} \\ \hline \end{array}$

15. $\begin{array}{r} 3\frac{1}{3} \\ + 5\frac{2}{3} \\ \hline \end{array}$

16. $\begin{array}{r} 1\frac{4}{5} \\ + 5\frac{3}{5} \\ \hline \end{array}$

Subtracting Mixed Numbers with Like Denominators

To subtract mixed numbers, subtract the whole numbers and fractions separately:

$$5\frac{5}{7} - 3\frac{1}{7} = (5 - 3) + (\frac{5}{7} - \frac{1}{7}) = 2\frac{4}{7}$$

OR

Write the mixed numbers as improper fractions and subtract. Then, change the answer back into a mixed number:

$$5\frac{5}{7} - 3\frac{1}{7} = \frac{40}{7} - \frac{22}{7} = \frac{18}{7} = 2\frac{4}{7}$$

Solve. Subtract the whole numbers and fractions separately.

1. $5\frac{5}{8} - 3\frac{4}{8} =$ (_____ – _____) + (— – —) = _____

2. $7\frac{3}{5} - 5\frac{1}{5} =$ (_____ – _____) + (— – —) = _____

3. $2\frac{1}{6} - 1\frac{1}{6} =$ (_____ – _____) + (— – —) = _____

4. $4\frac{9}{10} - 2\frac{7}{10} =$ (_____ – _____) + (— – —) = _____

Solve. Write the mixed numbers as improper fractions and subtract. Show the answer as a mixed number in simplest form.

5. $7\frac{2}{3} - 3\frac{1}{3} =$ _____ – _____ = _____ = _____

6. $6\frac{7}{8} - 1\frac{1}{8} =$ _____ – _____ = _____ = _____

7. $8\frac{4}{10} - 4\frac{2}{10} =$ _____ – _____ = _____ = _____

8. $9\frac{6}{7} - 2\frac{4}{7} =$ _____ – _____ = _____ = _____

Subtracting Mixed Numbers with Like Denominators

If the numerator of the first number is smaller than the numerator of the second number, borrow from the whole number. Add the whole to the fraction to create an improper fraction. Then, subtract.

$$6\frac{2}{5}$$
$$-3\frac{4}{5}$$

$\longrightarrow$

$$5\frac{7}{5}$$
$$-3\frac{4}{5}$$
$$2\frac{3}{5}$$

Borrow 1 from the 6 as $\frac{5}{5}$ and add to $\frac{2}{5}$.

Solve each problem. Write the answer in simplest form.

1.
$$5\frac{5}{8}$$
$$-2\frac{4}{8}$$

2.
$$6\frac{2}{6}$$
$$-3\frac{5}{6}$$

3.
$$5\frac{5}{8}$$
$$-3\frac{4}{8}$$

4.
$$7\frac{3}{5}$$
$$-5\frac{4}{5}$$

5.
$$8\frac{4}{5}$$
$$-4\frac{1}{5}$$

6.
$$3\frac{2}{6}$$
$$-2\frac{1}{6}$$

7.
$$9\frac{3}{7}$$
$$-2\frac{5}{7}$$

8.
$$8\frac{5}{9}$$
$$-3\frac{6}{9}$$

9.
$$5\frac{1}{3}$$
$$-1\frac{2}{3}$$

10.
$$10\frac{1}{4}$$
$$-7\frac{3}{4}$$

11.
$$5\frac{3}{3}$$
$$-4\frac{2}{3}$$

12.
$$4\frac{9}{10}$$
$$-2\frac{7}{10}$$

Subtracting Mixed Numbers with Like Denominators

Solve each problem. Write the answer in simplest form.

1. $12\frac{7}{8}$
   $-5\frac{5}{8}$

2. $2\frac{2}{3}$
   $-2\frac{1}{3}$

3. $9\frac{7}{8}$
   $-4\frac{4}{8}$

4. $3\frac{1}{8}$
   $-1\frac{7}{8}$

5. $10\frac{1}{5}$
   $-7\frac{4}{5}$

6. $3\frac{1}{4}$
   $-2\frac{3}{4}$

7. $10\frac{2}{3}$
   $-9\frac{1}{3}$

8. $5\frac{4}{5}$
   $-4\frac{1}{5}$

9. $5\frac{1}{3}$
   $-1\frac{2}{3}$

10. $8\frac{7}{10}$
    $-7\frac{9}{10}$

11. $8\frac{3}{16}$
    $-7\frac{5}{16}$

12. $6\frac{7}{15}$
    $-2\frac{8}{15}$

13. $6\frac{2}{12}$
    $-3\frac{2}{12}$

14. $4\frac{5}{6}$
    $-2\frac{1}{6}$

15. $4\frac{11}{18}$
    $-1\frac{7}{18}$

16. $8\frac{3}{10}$
    $-1\frac{7}{10}$

Multiplying Fractions by Whole Numbers

Fractions can be used to identify part of a set.

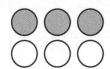

There are 6 circles. One-half ($\frac{1}{2}$) of the circles are shaded.

$\frac{1}{2}$ of 6 = 3

If you do not have a picture to find a fraction of a set, use division to help.

To find $\frac{1}{2}$ of 6, divide the whole number, 6, by the denominator, 2.
Multiply the quotient, 3, by the numerator, 1.

$$6 \div 2 = 3 \qquad 3 \times 1 = 3 \qquad \frac{1}{2} \text{ of } 6 = 3$$

To find $\frac{2}{3}$ of 24, divide the whole number, 24, by the denominator, 3.
Multiply the quotient, 8, by the numerator, 2.

$$24 \div 3 = 8 \qquad 8 \times 2 = 16 \qquad \frac{2}{3} \text{ of } 24 = 16$$

Solve each problem.

1. $\frac{2}{11}$ of 44 =

2. $\frac{2}{7}$ of 49 =

3. $\frac{2}{4}$ of 36 =

4. $\frac{4}{6}$ of 48 =

5. $\frac{3}{9}$ of 81 =

6. $\frac{3}{12}$ of 24 =

7. $\frac{2}{3}$ of 33 =

8. $\frac{3}{4}$ of 20 =

9. $\frac{1}{3}$ of 12 =

Multiplying Fractions by Whole Numbers

To multiply fractions, change the whole number to a fraction with a denominator of 1. Then, multiply the numerators and multiply the denominators. Last, simplify.

Solve each problem. Write the answer in simplest form.

1. $4 \times \dfrac{1}{2} =$

2. $2 \times \dfrac{2}{5} =$

3. $4 \times \dfrac{2}{7} =$

4. $3 \times \dfrac{5}{6} =$

5. $8 \times \dfrac{1}{8} =$

6. $\dfrac{2}{15} \times 3 =$

7. $\dfrac{1}{8} \times 5 =$

8. $\dfrac{5}{7} \times 5 =$

9. $\dfrac{2}{3} \times 2 =$

10. $\dfrac{3}{16} \times 4 =$

11. $\dfrac{1}{3} \times 7 =$

12. $4 \times \dfrac{3}{4} =$

13. $\dfrac{6}{8} \times 2 =$

14. $5 \times \dfrac{4}{5} =$

15. $3 \times \dfrac{2}{3} =$

Multiplying Fractions by Whole Numbers

Solve each problem. Write the answer in simplest form.

1. $9 \times \dfrac{2}{3} =$

2. $14 \times \dfrac{4}{7} =$

3. $77 \times \dfrac{10}{11} =$

4. $36 \times \dfrac{2}{288} =$

5. $16 \times \dfrac{4}{8} =$

6. $9 \times \dfrac{5}{6} =$

7. $3 \times \dfrac{1}{3} =$

8. $30 \times \dfrac{3}{90} =$

9. $12 \times \dfrac{1}{36} =$

10. $5 \times \dfrac{2}{5} =$

11. $12 \times \dfrac{7}{8} =$

12. $5 \times \dfrac{3}{40} =$

13. $22 \times \dfrac{1}{44} =$

14. $4 \times \dfrac{1}{8} =$

15. $81 \times \dfrac{2}{3} =$

Relating Decimals and Fractions

A **decimal** is a number that uses a **decimal point (.)** to show tenths and hundredths instead of a fraction.

A **tenth** is one out of 10 equal parts of a whole.

A **hundredth** is one out of 100 equal parts of a whole.

 $1\frac{6}{10}$

1.6

1 or 1.0

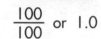 $\frac{6}{10}$ or 0.6

$\frac{100}{100}$ or 1.0

 $1\frac{47}{100}$

1.47

$\frac{47}{100}$ or 0.47

Write two ways to name each picture.

      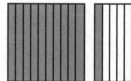

1. Fraction: _____

   Decimal: _____

2. Fraction: _____

   Decimal: _____

3. Fraction: _____

   Decimal: _____

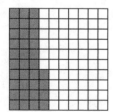   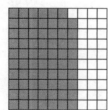      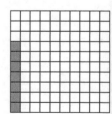

4. Fraction: _____

   Decimal: _____

5. Fraction: _____

   Decimal: _____

6. Fraction: _____

   Decimal: _____

Relating Decimals and Fractions

Write each decimal as a fraction in simplest form.

1. $0.5 =$ 2. $0.9 =$ 3. $0.7 =$

4. $9.5 =$ 5. $1.8 =$ 6. $2.2 =$

7. $0.62 =$ 8. $1.25 =$ 9. $0.1 =$

10. $0.22 =$ 11. $4.10 =$ 12. $0.36 =$

Write each fraction as a decimal.

13. $\frac{3}{10} =$ 14. $\frac{9}{10} =$ 15. $8\frac{8}{10} =$

16. $2\frac{5}{10} =$ 17. $\frac{4}{10} =$ 18. $\frac{88}{100} =$

19. $\frac{52}{100} =$ 20. $3\frac{25}{100} =$ 21. $6\frac{5}{10} =$

22. $\frac{14}{100} =$ 23. $4\frac{1}{10} =$ 24. $9\frac{30}{100} =$

Relating Decimals and Fractions

Write each as a decimal or a fraction in simplest form.

1. 8.2 =

2. 5.4 =

3. $48\frac{2}{10}$ =

4. 0.15 =

5. 25.32 =

6. 3.25 =

7. 30.2 =

8. $\frac{65}{100}$ =

9. 9.1 =

10. 10.6 =

11. $\frac{20}{100}$ =

12. $\frac{2}{10}$ =

13. 86.12 =

14. 6.5 =

15. $9\frac{9}{10}$ =

16. $1\frac{29}{100}$ =

17. 7.6 =

18. $\frac{99}{100}$ =

19. 0.75 =

20. 4.36 =

21. 9.45 =

22. $75\frac{2}{100}$ =

23. 25.2 =

24. $20\frac{6}{10}$ =

Comparing Decimals

> To compare decimals, first look at the whole numbers.
>
> $\underline{2}.68 < \underline{4}.43$ $\underline{48}.52 > \underline{12}.71$
>
> If the whole numbers are the same, compare the tenths.
>
> $7.\underline{8}3 > 7.\underline{3}8$ $80.\underline{7}4 > 80.\underline{0}7$
>
> If the whole number and the tenths are the same, compare the hundredths.
>
> $54.1\underline{9} > 54.1\underline{2}$ $3.4\underline{0} < 3.4\underline{4}$

Use > or < to compare the decimals.

1. 9.4 ◯ 9.9

2. 4.7 ◯ 4.2

3. 34.93 ◯ 49.04

4. 8.68 ◯ 8.30

5. 2.8 ◯ 2.3

6. 17.45 ◯ 17.46

7. 22.31 ◯ 22.18

8. 1.4 ◯ 1.7

9. 5.6 ◯ 5.8

10. 35.73 ◯ 35.81

11. 11.1 ◯ 1.11

12. 3.97 ◯ 3.99

13. 9.14 ◯ 91.4

14. 35.1 ◯ 55.9

15. 6.4 ◯ 4.9

Comparing Decimals

When comparing decimals, first compare the whole numbers. Then, compare the digits in the tenths columns. If the digits in the tenths columns are the same, compare the digits in the hundredths columns.

Use > or < to compare the decimals.

1.   0.6 ◯ 0.4   2.   0.1 ◯ 0.5   3.   0.23 ◯ 0.03   4.   0.6 ◯ 0.9

5.   0.06 ◯ 0.60   6.   0.4 ◯ 0.7   7.   0.9 ◯ 0.5   8.   0.7 ◯ 0.6

9.   0.42 ◯ 0.14   10.   0.72 ◯ 0.27   11.   0.25 ◯ 0.52   12.   0.7 ◯ 0.3

13.   1.4 ◯ 1.6   14.   3.5 ◯ 3.7   15.   16.2 ◯ 16.8   16.   5.21 ◯ 5.38

17.   2.48 ◯ 2.35   18.   14.5 ◯ 14.3   19.   42.6 ◯ 42.3   20.   3.8 ◯ 3.9

Comparing Decimals

Use >, <, or = to compare the decimals.

1. 8.0 ◯ 0.8   2. 2.51 ◯ 2.5   3. 6.62 ◯ 6.67   4. 0.99 ◯ 0.99

5. 0.7 ◯ 0.77   6. 4.3 ◯ 4.31   7. 1.4 ◯ 0.14   8. 2.04 ◯ 2.05

9. 5.3 ◯ 5.27   10. 8.75 ◯ 8.73   11. 71.05 ◯ 72.05   12. 3.0 ◯ 0.30

13. 9.2 ◯ 9.20   14. 43.1 ◯ 4.31   15. 6.1 ◯ 6.13   16. 18.9 ◯ 1.89

17. 0.20 ◯ 2.0   18. 3.41 ◯ 3.14   19. 6.06 ◯ 6.60   20. 281.01 ◯ 281

21. 0.50 ◯ 0.5   22. 7.38 ◯ 73.8   23. 0.01 ◯ 0.10   24. 54.2 ◯ 45.20

Customary Measurement

US customary units of length

12 inches (in.) = 1 foot (ft.)

3 feet (ft.) = 1 yard (yd.)

5,280 feet (ft.) = 1 mile (mi.)

1,760 yards (yd.) = 1 mile (mi.)

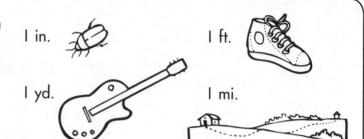

1 in.　　　　1 ft.

1 yd.　　　　1 mi.

US customary units of capacity and weight

2 cups (c.) = 1 pint (pt.)

2 pints = 1 quart (qt.)

4 quarts = 1 gallon (gal.)

16 ounces (oz.) = 1 pound (lb.)

2,000 pounds = 1 ton (t.)

1 oz.　　　1 lb.　　　1 t.

Circle the most appropriate unit of measure.

1.

　　in.　　　　yd.

2.

　　in.　　　　yd.

3.

　　yd.　　　　mi.

4.

　　mi.　　　　ft.

5.

　　ft.　　　　yd.

6.

　　in.　　　　ft.

7.

　　oz.　　　　lb.

8.

　　lb.　　　　t.

9.

　　oz.　　　　lb.

10.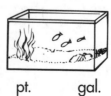

　　c.　　　　qt.

11.

　　pt.　　　　gal.

12.

　　oz.　　　　lb.

Customary Measurement

Remember:

12 inches (in.) = 1 foot (ft.)   16 ounces (oz.) = 1 pound (lb.)   2 cups (c.) = 1 pint (pt.)

3 feet = 1 yard (yd.)   2,000 pounds = 1 ton (t.)   2 pints = 1 quart (qt.)

5,280 feet = 1 mile (mi.)   4 quarts = 1 gallon (gal.)

1,760 yards = 1 mile

Use >, <, or = to compare the measurements.

1. 14 oz. ◯ 1 lb.

2. 4 c. ◯ 1 pt.

3. 2 qt. ◯ 2 pt.

4. 3 gal. ◯ 12 qt.

5. 1 qt. ◯ 3 pt.

6. 3 lb. ◯ 32 oz.

7. 2 t. ◯ 3,000 lb.

8. 4 c. ◯ 2 pt.

9. 1 t. ◯ 2,000 lb.

10. 2 c. ◯ 2 pt.

11. 3 qt. ◯ 1 gal.

12. 2 lb. ◯ 30 oz.

13. 8 qt. ◯ 3 gal.

14. 17 oz. ◯ 1 lb.

15. 1 qt. ◯ 4 pt.

16. 6 ft. ◯ 3 yd.

17. 2 ft. ◯ 20 in.

18. 1 yd. ◯ 36 in.

19. 1 mi. ◯ 100 yd.

20. 48 in. ◯ 4 ft.

21. 16,000 oz. ◯ 1 t.

22. 3 mi. ◯ 10,000 ft.

23. 10 in. ◯ 1 ft.

24. 5 yd. ◯ 50 ft.

Customary Measurement

Give the equivalent for each measurement.

1. 4 ft. = _____ in.

2. 84 in. = _____ ft.

3. 33 yd. = _____ ft.

4. 36 in. = _____ ft.

5. 48 ft. = _____ yd.

6. 4 lb. = _____ oz.

7. 1,200 oz. = _____ lb.

8. 6,000 lb. = _____ t.

9. 4.5 lb. = _____ oz.

10. 96 oz. = _____ lb.

11. 12 c. = _____ pt.

12. 3 gal.= _____ pt.

13. 2 pt. = _____ c.

14. 8 qt. = _____ pt.

15. 8 qt. = _____ c.

Solve each problem.

16. Brian needs 108 inches of pipe. How many feet of pipe does he need to buy?

17. Tess has 180 inches of ribbon. She uses 36 inches. How many yards of ribbon does she have left?

18. A produce truck that carries apples and oranges weighs 4 tons. How much does the truck weigh in pounds?

19. Meredith lifts two 5-pound weights every day. How many total ounces does she lift?

20. If Lindsay has 2 gallons of milk, how many pints does she have?

21. Pablo is making orange juice. If he has 8 quarts of juice, how many 1-cup servings can he pour?

Name _____

Metric Measurement

Metric units of length

100 centimeters (cm) = 1 meter (m)

1,000 meters = 1 kilometer (km)

Metric units of mass

1,000 grams (g) = 1 kilogram (kg)

Metric units of capacity

1,000 milliliters (mL) = 1 liter (L)

 1 centimeter (cm)

1 meter (m)

1 kilometer (km)

1 gram (g)

1 kilogram (kg)

1 milliliter (mL)

1 liter (L)

Circle the most appropriate measurement.

1.
1.5 km   1.5 m

2.
11 g   11 kg

3.
7 g   7 kg

4.
15 g   15 kg

5.
30 g   30 kg

6.
20 mm   20 cm

7.
5 mL   5 L

8.
25 cm   25 m

9.
400 mL   400 L

10.
255 mL   255 L

11.
16 cm   16 m

12.
800 mL   800 L

Metric Measurement

Use >, <, or = to compare the measurements.

1. 7 g ◯ 698 mg

2. 56 cm ◯ 6 m

3. 1,500 mL ◯ 1.5 L

4. 599 cm ◯ 5 m

5. 43 mg ◯ 5 g

6. 35 m ◯ 35 cm

7. 9,000 g ◯ 9 kg

8. 800 m ◯ 8 km

9. 3 L ◯ 2,000 mL

10. 100 cm ◯ 1 km

11. 30 kg ◯ 3,000 g

12. 2,500 mL ◯ 2 L

13. 600 m ◯ 6 cm

14. 9,000 mg ◯ 9 g

15. 750 L ◯ 7 mL

16. 3 km ◯ 3,200 m

17. 2.5 L ◯ 3,000 mL

18. 4,500 g ◯ 45 kg

19. 19 kg ◯ 1,900 g

20. 25,000 mg ◯ 2.5 g

Metric Measurement

Give the equivalent for each measurement.

1. 700 cm = _____ m   2. 8,000 m = _____ km   3. 15 m = _____ cm

4. 17 km = _____ m   5. 300,000 cm = _____ km   6. 3 g = _____ mg

7. 8,000 mg = _____ g   8. 650,000 mg = _____ kg   9. 0.8 kg = _____ mg

10. 12 g = _____ mg   11. 8 L = _____ mL   12. 5,000 mL = _____ L

13. 48,000 mL = _____ L   14. 0.4 L = _____ mL   15. 27 L = _____ mL

Solve each problem.

16. Penny walked 2 kilometers. Anita walked 5,000 meters. How many more meters did Anita walk than Penny?

17. Norman has a piece of string that measures 15 centimeters. Kayla has a piece of string that measures 200 millimeters. Who has the longer piece of string? How much longer is it?

18. Megan uses 4,000 milligrams of sugar in her recipe. How many grams of sugar does she use?

19. Harry measures 15 grams of salt. How many milligrams does he measure?

20. Karen drinks 0.5 liter of soft drink. How many milliliters does she drink?

21. Isabelle buys fifteen 2-liter bottles of soft drink for the party. Her guests drink 18,000 milliliters. How many liters of soft drink does Isabelle have left?

Finding Perimeter

Perimeter is the total distance around a given figure. To find the perimeter, add the lengths of the sides of the figure.

Example:

Perimeter = 4 cm + 8 cm + 4 cm + 8 cm

P = 24 cm

8 cm

4 cm   4 cm

8 cm

Find the perimeter.

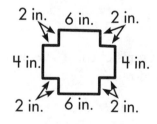

15 yd.

10 yd.   10 yd.

15 yd.

1.   P = _____

8 in.

6 in.

7 in.

2.   P = _____

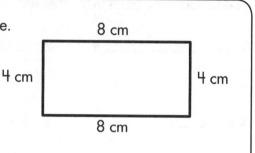

2 yd.
8 yd.   4 yd.

4 yd.

8 yd.   4 yd.
2 yd.

3.   P = _____

3 cm   3 cm

3 cm   3 cm

4.   P = _____

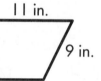

11 in.

9 in.   9 in.

11 in.

5.   P = _____

12 ft.   12 ft.

12 ft.

6.   P = _____

2 in.   6 in.   2 in.

4 in.   4 in.

2 in.   6 in.   2 in.

7.   P = _____

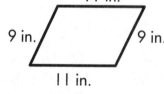

4 cm

3 cm   3 cm

4 cm

8.   P = _____

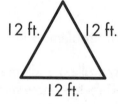

18 mm

9 mm   9 mm

14 mm

9.   P = _____

13 mm
                7 mm
9 mm
                7 mm
13 mm

10. P = _____

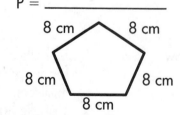

8 cm   8 cm

8 cm   8 cm

8 cm

11. P = _____

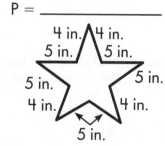

4 in. 4 in.
5 in.   5 in.

5 in.   5 in.

4 in.   4 in.

5 in.

12. P = _____

Finding Perimeter

Use the formula P = (2 × l) + (2 × w) to find the perimeter of each quadrilateral.

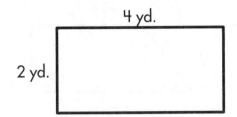

4 yd.

2 yd.

1. P = _____

6 ft.

3 ft.

2. P = _____

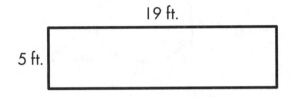

19 ft.

5 ft.

3. P = _____

6 in.

7 in.

4. P = _____

8 in.

8 in.

5. P = _____

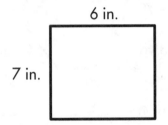

12 mm

9 mm

6. P = _____

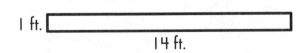

1 ft.

14 ft.

7. P = _____

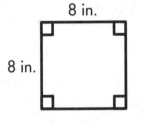

9 mm

3 mm

8. P = _____

Finding Perimeter

Use the information given to find the missing side length of each quadrilateral. Show your work.

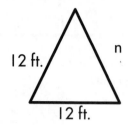

1. P = 36 ft., n = _____

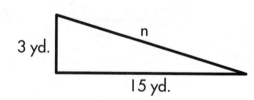

2. P = 68 cm, n = _____

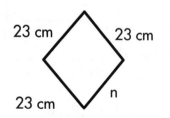

3. P = 21 yd., n = _____

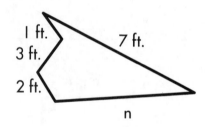

4. P = 18 in., n = _____

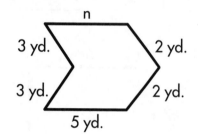

5. P = 78 mm, n = _____

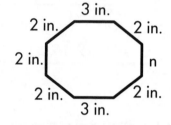

6. P = 19 ft., n = _____

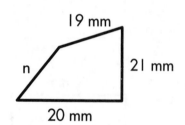

7. P = 92 cm, n = _____

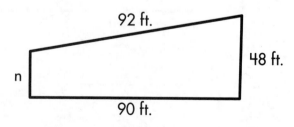

8. P = 36 yd., n = _____

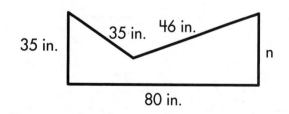

9. P = 231 in., n = _____

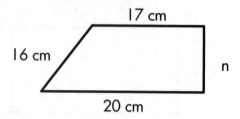

10. P = 270 ft., n = _____

Finding Area

Area is the number of square units enclosed within a boundary. Area is measured in different units such as square feet or square centimeters. For example, 14 square units are in this figure.

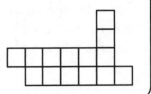

Find the area of each figure.

1.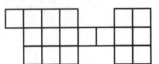

_____ square units

2.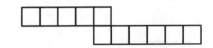

_____ square units

3.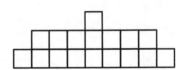

_____ square units

4.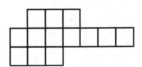

_____ square units

5.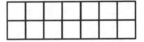

_____ square units

6.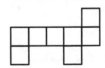

_____ square units

7.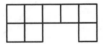

_____ square units

8.

_____ square units

9.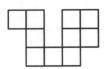

_____ square units

10.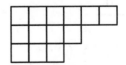

_____ square units

Name _____

Finding Area

The area of a figure tells how many square units are needed to cover the figure. Area can be measured in different units, such as square feet, square meters, or square inches. Use the following formulas to find the area of a square or a rectangle.

Square A = s × s   Rectangle A = lw

Find the area of each figure.

1.
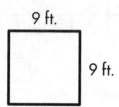
9 ft.

9 ft.

A = _____

2.
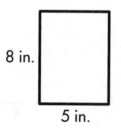
8 in.

5 in.

A = _____

3.
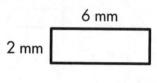
6 mm

2 mm

A = _____

4.
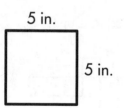
21 cm

8 cm

A = _____

5.
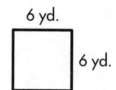
6 yd.

6 yd.

A = _____

6.
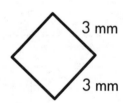
19 ft.

4 ft.

A = _____

7.
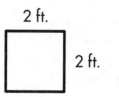
5 in.

5 in.

A = _____

8.
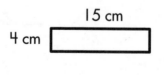
15 cm

4 cm

A = _____

9.
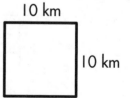
3 mm

3 mm

A = _____

10.
2 ft.

2 ft.

A = _____

11.
13 m

7 m

A = _____

12.
10 km

10 km

A = _____

© Carson-Dellosa • CD-104629

Finding Area

Find the area of each figure. Remember to label the units.

1. A = _____

6 mi.

6 mi.

2. A = _____

7 in.

6 in.

3. A = _____

4 yd.

3 yd.

4. A = _____

5 m

6 m

Find the area of each quadrilateral with the given dimensions.

	Length	**Width**	**Area**
5.	10 in.	6 in.	
6.	5 cm	3 cm	
7.	4 yd.	3 yd.	
8.	10 km	4 km	
9.	4 mi.	4 mi.	
10.	6 ft.	3 ft.	

Line Plots

A **line plot** is a type of graph that shows information on a number line.

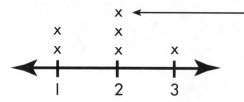

Line plots are useful for showing frequency, or the number of times something is repeated.

1. Use a ruler to measure 8 things to the nearest $\frac{1}{2}$ inch. Record your data on the table.

Item	Length	Item	Length

2. Use the data from the table to make a line plot.

- First, look at the data and decide what numbers you will need to include.

- Then, mark each number on the line plot and label it. Do not leave out numbers in between, even if they have no data!

- Finally, mark an X on the line plot to represent each piece of data.

Line Plots

1. Use a ruler to measure 10 things to the nearest $\frac{1}{4}$ inch. Record your data on the table.

Item	Length	Item	Length

2. Use the data from the table to make a line plot. Remember to look at your data to see what numbers you need to represent. Then, divide and label the line. Mark each data point with an X.

Line Plots

1. Use a ruler to measure 10 things to the nearest $\frac{1}{8}$ inch. Record your data on the table.

Item	Length	Item	Length

2. Use the data from the table to make a line plot.

Identifying and Measuring Angles

Using a protractor will help you draw and measure angles accurately.

How to Use a Protractor

1. Find the center dot or intersecting segments along the straight edge on the bottom of the protractor.

2. Place the dot or intersecting segments over the vertex, or point, of the angle you wish to measure.

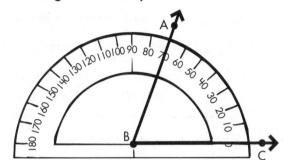

3. Rotate the protractor so that the zero mark on the straight edge lines up with one side of the angle.

4. Determine which set of numbers you will use. Find the point where the second side of the angle intersects the numbered edge of the protractor.

5. Read the number that is written on the protractor at the point of intersection. This is the measure of the angle in degrees.

Use a protractor to measure the angle to the nearest degree. Write the number of degrees and the type of angle (*right*, *obtuse*, or *acute*).

1.

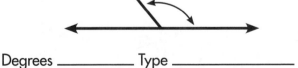

Degrees _____ Type _____

2.

Degrees _____ Type _____

3.

Degrees _____ Type _____

4.

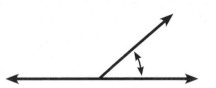

Degrees _____ Type _____

5.

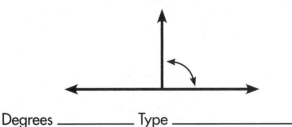

Degrees _____ Type _____

6.

Degrees _____ Type _____

Identifying and Measuring Angles

When measuring angles, remember:

1. Use a ruler to extend the rays if they are too short to measure.

2. Align the base ray on the protractor's straight edge (0°).

3. Read the number the ray intersects. Use the type of angle (*acute* or *obtuse*) to help you decide which set of numbers to use.

Measure each angle with a protractor.

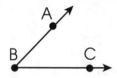

1. _____

2. _____

3. _____

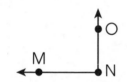

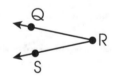

4. _____

5. _____

6. _____

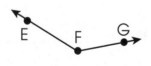

7. _____

8. _____

9. _____

Identifying and Measuring Angles

Use a protractor to measure each specified angle to the nearest degree. Write the type (*right*, *obtuse*, or *acute*) and measure of each angle.

1.

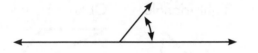

Degrees _____ Type _____

2.

Degrees _____ Type _____

3.

Degrees _____ Type _____

4.

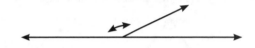

Degrees _____ Type _____

5.

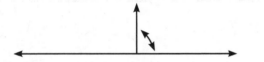

Degrees _____ Type _____

6.

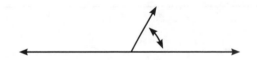

Degrees _____ Type _____

7.

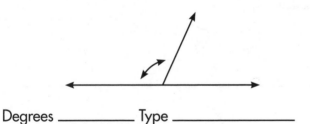

Degrees _____ Type _____

8.

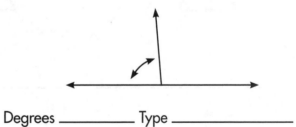

Degrees _____ Type _____

Use a protractor to draw each angle.

9. 65°

10. 130°

Identifying Triangles by Angle

A **triangle** is a three-sided polygon. A triangle's angles can be used to classify it.

Acute Triangle	Equiangular Triangle	Right Triangle	Obtuse Triangle
		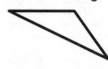	
three acute angles	three congruent angles	one right angle	one obtuse angle

Classify each triangle below by its angles. Write *acute, equiangular, right,* or *obtuse.*

1.

_____

2.

_____

3.

_____

4.

_____

5.

_____

6.

_____

7.
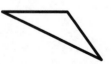

_____

8.

_____

9.

_____

Identifying Triangles by Angle

The sum of the three angles of every triangle equals 180°.

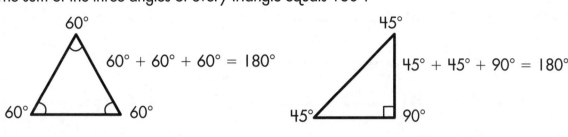

$60° + 60° + 60° = 180°$

$45° + 45° + 90° = 180°$

Below are sets of three angles from various types of triangles. Match each set of angles to the most appropriate term.

1. angles 60°, 60°, 60° _____

2. angles 30°, 60°, 90° _____

3. angles 89°, 56°, 35° _____

4. angles 40°, 40°, 100° _____

5. angles 30°, 50°, 100° _____

6. angles 90°, 45°, 45° _____

A. acute

B. obtuse

C. right

D. equiangular

Circle **T** for True or **F** for False.

7. A triangle with angles of 110°, 30°, and 40° is obtuse.   T   F

8. A triangle with angles of 60°, 40°, and 80° is right.   T   F

9. A triangle with angles of 60°, 60°, and 60° is equiangular.   T   F

10. A triangle with angles of 70°, 50°, and 60° is acute.   T   F

Identifying Triangles by Angle

Match each description with the correct triangle names from the box.

1. Angles measure 90°, 100°, 30°. _____

2. Angles measure 30°, 60°, 90°. _____

3. Angles measure 25°, 10°, 145°. _____

4. Angles measure 15°, 30°, 12°. _____

5. Angles measure 45°, 90°, 45°. _____

6. Angles measure 60°, 60°, 60°. _____

7. Angles measure 50°, 88°, 42°. _____

| A. right |
| B. obtuse |
| C. acute |
| D. equiangular |
| E. not a possible triangle |

Complete each statement using *sometimes, always,* or *never.*

8. An obtuse triangle is _____ an equilateral triangle.

9. A right triangle _____ has a right angle and two acute angles.

10. An acute triangle is _____ an equiangular triangle.

Explain why each shape below is not possible.

11. Right obtuse triangle

_____

_____

12. Triangle with two obtuse angles

_____

_____

Identifying Lines, Rays, and Line Segments

A **ray** is a portion of a line that extends from one endpoint infinitely in one direction. The ray to the right is named $\overrightarrow{AB}$, with the endpoint written first and any point on the ray written next.

A **line segment** is a finite portion of a line that contains two endpoints. The segment to the right is named $\overline{AB}$. The segment must be named by its two endpoints.

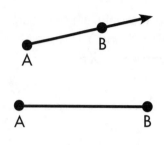

Identify the following as a *line, ray, line segment,* or *points*.

1.

_____

2.

_____

3.

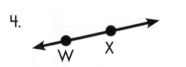

_____

4.

_____

5.

_____

6.

_____

7.

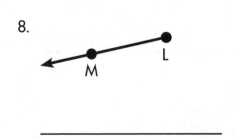

_____

8.

_____

Identifying Lines, Rays, and Line Segments

The straight path between points X and Y is a **line segment**. ($\overline{XY}$)

A **line** is a straight path that goes unending in two directions. ($\overleftrightarrow{CD}$)

A **ray** is a straight path that begins at a point and goes unending in one direction. ($\overrightarrow{TX}$)

Lines that never meet are called **parallel lines**.

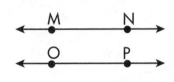

Lines that cross are called **intersecting lines**.

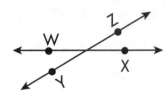

Lines that cross at right angles are called **perpendicular lines**.

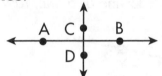

Identify each as a *line segment, line,* or *ray.* Use letters to name it.

1. _____   2. _____   3. _____   4. _____

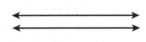

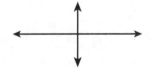

5. _____   6. _____   7. _____   8. _____

Identify each as *parallel, intersecting,* or *perpendicular* lines.

9. _____   10. _____   11. _____   12. _____

13. Draw two parallel lines.
    Draw two intersecting lines
    across the two parallel lines.

Identifying Lines, Rays, and Line Segments

Draw and label each of the following.

1. $\overrightarrow{AB}$

2. Points C and D

3. $\overline{RS}$

4. Parellel lines $\overleftrightarrow{LM}$ and $\overleftrightarrow{NO}$

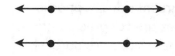

5. Perpendicular lines $\overleftrightarrow{MN}$ and $\overleftrightarrow{LM}$

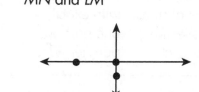

6. $\overrightarrow{JK}$

Use the figure to the right to answer each question.

7. Name four points. _____

8. Name two parallel line segments. _____

9. Name two intersecting lines. _____

10. Name three rays. _____

Use the figure below to answer each question.

11. Name three points. _____

12. Name the two perpendicular lines. _____

13. Name four line segments. _____

14. Name four rays. _____

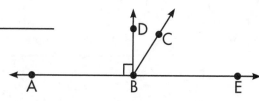

Identifying Polygons

A **parallelogram** is a quadrilateral with opposite sides parallel. Opposite sides and angles are congruent, or equal in length.

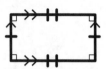 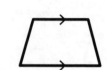

A **rectangle** is a parallelogram with four right angles. Opposite sides are congruent and parallel.

A **square** is a rectangle with congruent sides. Opposite sides are parallel.

A **trapezoid** is a quadrilateral with exactly one pair of parallel sides.

A **rhombus** is a parallelogram with four congruent sides. Opposite angles are congruent and opposite sides are parallel.

Name each figure.

1.

_____

2.

_____

3.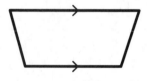

_____

4.

_____

5.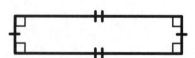

_____

6.

_____

7.

_____

8.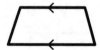

_____

Identifying Polygons

Identify each figure by writing its name and a brief description of the figure.

	Name	Description

1. 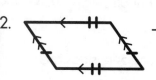 _____ _____
_____
_____

2. _____ _____
_____
_____

3.  _____ _____
_____
_____

4. _____ _____
_____
_____

5. _____ _____
_____
_____

6. _____ _____
_____
_____

7. _____ _____
_____
_____

8. _____ _____
_____
_____

Identifying Polygons

Draw a line to match the name of each polygon to its definition.

1. pentagon

2. parallelogram

3. rhombus

4. hexagon

5. rectangle

6. octagon

7. triangle

8. trapezoid

9. quadrilateral

10. polygon

A. a polygon with six sides

B. a quadrilateral with opposite sides that are parallel and congruent

C. a polygon with three angles

D. a quadrilateral in which all angles are right angles—opposite sides are parallel and congruent

E. a rectangle with all four sides congruent—opposite sides are parallel, and opposite angles are congruent

F. a polygon with five sides

G. three or more line segments connected so that an area is closed in

H. a polygon with eight sides

I. any polygon with four sides

J. a quadrilateral with two nonparallel sides and two parallel sides

Draw each figure on the dot grid.

11. Draw a triangle.

12. Draw a pentagon.

13. Draw a quadrilateral.

14. Draw an octagon.

15. Draw a trapezoid.

16. Draw a hexagon.

Exploring Line Symmetry

A figure has **line symmetry** if it can be folded along a line so that the two halves are mirror images.

These figures have line symmetry. The heart has one line of symmetry. The rectangle has two lines of symmetry.

 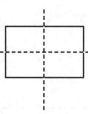

These figures do not have line symmetry.

Determine if the following figures have line symmetry and write *yes* or *no*. If *yes*, draw all of the lines of symmetry.

1.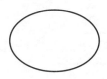

_____

2.

_____

3.

_____

4.

_____

5.

_____

6.

_____

7.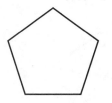

_____

8.

_____

9.

_____

Exploring Line Symmetry

Some objects have more than one line of symmetry. A regular hexagon has six lines of symmetry.

Decide whether these figures have one line of symmetry, two lines of symmetry, or no lines of symmetry. Write *one, two,* or *none.* Then, draw the line or lines.

1.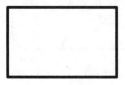

_____

2.

_____

3.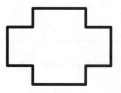

_____

4.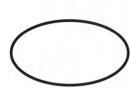

_____

5.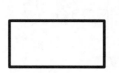

_____

6.

_____

7.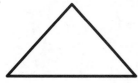

_____

8.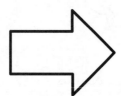

_____

9.

_____

10.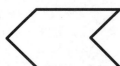

_____

11.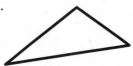

_____

12.

_____

Exploring Line Symmetry

Complete each design to show symmetry. Then, draw the line of symmetry.

1.

2.

3.

4.

5.

6.

7.

8.

9.

10.

11.

12.

13.

14.

15.

Answer Key

Name _____ (4.OA.A.2, 4.OA.A.3)

Solving Word Problems

Circle the correct problem to solve each problem. Solve.

1. There were 415 people who attended the school carnival. Of those, 316 rode the Ferris wheel. How many people did not ride the Ferris wheel?

$$\begin{array}{r} 415 \\ + 316 \end{array} \qquad \boxed{\begin{array}{r} 415 \\ - 316 \\ \hline 99 \end{array}}$$

2. At the carnival, 236 people ate pink cotton candy, and 178 people ate blue cotton candy. How many people in all ate cotton candy?

$$\boxed{\begin{array}{r} 236 \\ + 178 \\ \hline 414 \end{array}} \qquad \begin{array}{r} 236 \\ - 178 \end{array}$$

3. There were 23 rides at the carnival. If Whitney rode each ride 6 times, how many times did she ride in all?

$$\begin{array}{r} 23 \\ + 6 \end{array} \qquad \boxed{\begin{array}{r} 23 \\ \times 6 \\ \hline 138 \end{array}}$$

4. There were 160 kids who played relay races at the carnival. The kids were divided into 8 teams. How many kids were on each team?

$$\begin{array}{r} 160 \\ \times 8 \end{array} \qquad \boxed{8\overline{)160} \quad 20}$$

5. Carter sold 594 ride tickets in 1 hour at the carnival. If he sells the same amount each hour, how many tickets will he sell in 4 hours?

$$\boxed{\begin{array}{r} 594 \\ \times 4 \\ \hline 2{,}376 \end{array}} \qquad 4\overline{)594}$$

6. If 137 girls and 159 boys attended the carnival, how many more boys attended than girls?

$$\begin{array}{r} 137 \\ + 159 \end{array} \qquad \boxed{\begin{array}{r} 159 \\ - 137 \\ \hline 22 \end{array}}$$

5

Name _____ (4.OA.A.2, 4.OA.A.3)

Solving Word Problems

Write one or more equations for each problem. Solve.

1. The raffle ticket fundraiser sold 2,453 tickets last year and 3,832 tickets last year. How many more tickets did they sell this year than last year?

3,832 − 2,453 = 1,379 tickets

2. Each student received 3 ticket books to sell raffle tickets. There were 50 tickets in each book. If Ella turned in 98 unsold tickets, how many tickets did she sell?

3 × 50 = 150, 150 − 98 = 52 tickets

3. The bake sale fundraiser sold five dozen chocolate chip cookies, nine dozen sugar cookies, and six dozen oatmeal cookies. How many cookies did the fundraiser sell?

**5 × 12 = 60, 9 × 12 = 108,
6 × 12 = 72, 60 + 108 + 72 = 240 cookies**

4. The bake sale made $832 on Friday and $1,276 on Saturday. How much money did the bake sale make in all?

1,276 + 832 = $2,108

5. The school decided to divide the profits from their fundraising between 9 classrooms in the school. They will put any leftover money towards a new welcome mat for the school. If the school raised a total of $4,749, how much money will go to each classroom? How much money will go to buy the new welcome mat?

**4,749 ÷ 9 = 527r6
Each classroom will get $527, and there will be $6 for the welcome mat.**

6. Kayla won the raffle for an afternoon at Ace Arcade. She gets 200 free tokens. If each game takes 3 tokens, how many games can she play? If she finds another token on the ground, can she play one more game? Explain.

**200 ÷ 3 = 66r2, 66 games
Yes, she had 2 tokens left, so now she has 3 tokens.**

6

Name _____ (4.OA.A.2, 4.OA.A.3)

Solving Word Problems

Solve each problem.

1. The fourth grade is going on a field trip to Colonial Town. There are three fourth grade classes, each with 19 students. There must be one chaperone for every 9 students. How many chaperones will need to go on the field trip?

7 chaperones

2. Colonial Town has an average of 7,895 total visitors on a weekend day and an average of 3,638 total visitors on a weekday. During the week, the average number of student visitors on field trips is 2,493. Not counting students on a field trip, how many more visitors on average are there on a weekend day than on a week day?

6,750 visitors

3. Students on a field trip to Colonial Town get to make their own candles. If the average number of students in a class is 23, and 38 classes of students have field trips each week, what is the average number of candles made by students each week?

874 candles

4. The teachers buy cookies from the bakery for the students. They want each of their 73 students to get 4 cookies. If the cookies come in packages of 9, how many more packages do they need to buy?

33 packages

5. At the blacksmith's shop, the students learn that the blacksmith forge gets as hot as 1400°F. How many times hotter is the forge than the typical air temperature of 70°F?

20 times hotter

6. The blacksmith tells the students that he and his apprentice have been working on making nails for building projects and repairs in the town. They made 964 nails the first week of the month, 1,072 nails the second week, 936 nails the third week, and 1,113 nails the fourth week of the month. They will bundle the nails in boxes of 100. How many boxes will they need?

41 boxes

7. Write a division word problem in which you would have to interpret the remainder.

Answers will vary.

_____

_____

7

Name _____ (4.OA.B.4, 4.NBT.B.5)

Determining Factors and Multiples

> **Factors** are numbers multiplied together. The first factor tells the number of sets. The second factor tells the number in each set.
>
> How many different factors can you name for one number?
>
> 2 × 3 = 6
> 6 × 1 = 6
> factors: 1, 2, 3, 6

Write all of the multiplication sentences for each set. Then, list the factors.

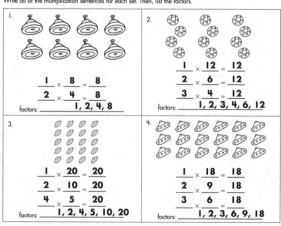

1.
$$\underline{1} \times \underline{8} = \underline{8}$$
$$\underline{2} \times \underline{4} = \underline{8}$$
factors: __1, 2, 4, 8__

2.
$$\underline{1} \times \underline{12} = \underline{12}$$
$$\underline{2} \times \underline{6} = \underline{12}$$
$$\underline{3} \times \underline{4} = \underline{12}$$
factors: __1, 2, 3, 4, 6, 12__

3.
$$\underline{1} \times \underline{20} = \underline{20}$$
$$\underline{2} \times \underline{10} = \underline{20}$$
$$\underline{4} \times \underline{5} = \underline{20}$$
factors: __1, 2, 4, 5, 10, 20__

4.
$$\underline{1} \times \underline{18} = \underline{18}$$
$$\underline{2} \times \underline{9} = \underline{18}$$
$$\underline{3} \times \underline{6} = \underline{18}$$
factors: __1, 2, 3, 6, 9, 18__

5. What number, other than 1, is a common factor for all 4 numbers in problems 1-4? __2__

So, we can say that all of these numbers are multiples of __2__.

8

Answer Key

Name _____   4.OA.B.4, 4.OA.C.5, 4.NBT.B.5

Determining Factors and Multiples

Multiples are numbers that all have the same particular factor.

Example: Even numbers are all multiples of the factor 2.

multiples:	2	4	6	8	10	12	14
other factor:	1	2	3	4	5	6	7

List the multiples of each factor. Write the other factor in the factor pair below each multiple.

1. multiples of the factor 5

multiples:	10	**15**	20	**25**	**30**	35	**40**	**45**
other factor:	**2**	**3**	**4**	**5**	**6**	**7**	**8**	**9**

2. multiples of the factor 4

multiples:	12	**16**	20	**24**	**28**	32	**36**	40
other factor:	**3**	**4**	**5**	**6**	**7**	**8**	**9**	**10**

3. multiples of the factor 3

multiples:	15	**18**	**21**	24	**27**	**30**	**33**	**36**
other factor:	**5**	**6**	**7**	**8**	**9**	**10**	**11**	**12**

4. multiples of the factor 6

multiples:	6	12	18	**24**	**30**	36	**42**	**48**
other factor:	**1**	**2**	**3**	**4**	**5**	**6**	**7**	**8**

5. multiples of the factor 9

multiples:	18	**27**	**36**	**45**	54	**63**	72	**81**
other factor:	**2**	**3**	**4**	**5**	**6**	**7**	**8**	**9**

6. multiples of the factor 8

multiples:	32	**40**	**48**	56	**64**	**72**	80	**88**
other factor:	**4**	**5**	**6**	**7**	**8**	**9**	**10**	**11**

9

Name _____   4.OA.B.4, 4.NBT.B.5

Determining Factors and Multiples

A **composite number** has factors other than 1 and itself.

A **prime number** only has the factors 1 and itself.

Name any factors, other than 1, that each group of composite numbers are multiples of.

1. 36, 27, 99, 45, 63: _____ **3, 9**

2. 40, 20, 70, 100, 60: _____ **2, 5, 10**

3. 8, 32, 12, 48, 20: _____ **2, 4**

4. 42, 21, 56, 84, 35: _____ **7**

5. 20, 10, 60, 40, 25: _____ **5**

Write 4 multiples that have both factors in common.

6. 2 and 3: _____ **6, 12, 18, 24**

7. 4 and 6: _____ **12, 24, 36, 48**

8. 5 and 2: _____ **10, 20, 30, 40**

Answers will vary but may include the numbers shown.

Write the factor pairs for each number. Tell whether it is a composite number or a prime number.

9. 36: **1, 36; 2, 18; 3, 12; 4, 9; 6, 6**   composite

10. 14: **1, 14; 2, 7**   composite

11. 23: **1, 23**   prime

12. 33: **1, 33; 3, 11**   composite

13. 48: **1, 48; 2, 24; 3, 16; 4, 12; 6, 8**   composite

14. 11: **1, 11**   prime

Name _____   4.NBT.A.1, 4.NBT.B.5, 4.NBT.B.6

Using Place Value to Multiply and Divide

Multiplying by 10s adds places to a number.

$200 \times 10 = 2,000$ hundreds become thousands

$50 \times 100 = 5,000$ tens become thousands

Dividing by 10s removes places from a number.

$700 \div 10 = 70$ hundreds become tens

$70,000 \div 100 = 700$ ten thousands become hundreds

Use place value to multiply or divide.

1. $800 \times 10 =$ **8,000**

2. $100,000 \div 10 =$ **10,000**

3. $30 \times 10 =$ **300**

4. $10 \times 100 =$ **1,000**

5. $9,000 \times 10 =$ **90,000**

6. $4 \times 100 =$ **400**

7. $50,000 \times 10 =$ **500,000**

8. $30 \times 100 =$ **3,000**

9. $4,000 \div 10 =$ **400**

10. $8,000 \times 100 =$ **800,000**

11. $70 \div 10 =$ **7**

12. $200 \div 100 =$ **2**

13. $20,000 \div 10 =$ **2,000**

14. $90,000 \div 100 =$ **900**

11

Name _____   4.NBT.A.1, 4.NBT.B.5, 4.NBT.B.6

Using Place Value to Multiply and Divide

Remember: When you multiply by a multiple of 10, add the number of zeros in the multiple of 10 to the first number:

$400 \text{ (2 zeros)} \times 100 \text{ (2 zeros)} = 40,000 \text{ (4 zeros)}$

When you divide by a multiple of 10, subtract the number of zeros in the multiple of 10 from the first number:

$40,000 \text{ (4 zeros)} \div 100 \text{ (2 zeros)} = 400 \text{ (2 zeros)}$

Use place value to multiply or divide.

1. $6,000 \times 10 =$ **60,000**

2. $60,000 \div 10 =$ **6,000**

3. $70,000 \times 10 =$ **700,000**

4. $4,000 \div 100 =$ **40**

5. $800 \times 100 =$ **80,000**

6. $700,000 \div 100 =$ **7,000**

7. $2,000 \times 100 =$ **200,000**

8. $500 \div 10 =$ **50**

9. $340 \times 10 =$ **3,400**

10. $280,000 \div 10 =$ **28,000**

11. $7,890 \times 10 =$ **78,900**

12. $79,000 \div 100 =$ **790**

13. $459 \times 1,000 =$ **459,000**

14. $5,240 \div 10 =$ **524**

15. $52 \times 10,000 =$ **520,000**

16. $582,000 \div 1,000 =$ **582**

17. $4 \times 100,000 =$ **400,000**

18. $36,000 \div 1,000 =$ **36**

19. $100,000 \times 10 =$ **1,000,000**

20. $900,000 \div 10,000 =$ **90**

Answer Key

(4.NBT.A.1, 4.NBT.B.5, 4.NBT.B.6)

Using Place Value to Multiply and Divide

Use place value to multiply or divide.

1. $5 \times 70 =$ **350**

2. $90 \times 2,000 =$ **180,000**

3. $4 \times 9,000 =$ **36,000**

4. $500 \div 5 =$ **100**

5. $200,000 \times 3 =$ **600,000**

6. $7,000 \div 70 =$ **100**

7. $50 \times 400 =$ **20,000**

8. $30,000 \div 300 =$ **100**

9. $30,000 \times 30 =$ **900,000**

10. $900 \div 90 =$ **10**

11. $3 \times 800 =$ **2,400**

12. $22,000 \div 2 =$ **11,000**

13. $9 \times 50,000 =$ **450,000**

14. $450 \div 90 =$ **5**

15. $600 \times 900 =$ **540,000**

16. $600 \div 30 =$ **20**

17. $500 \times 300 =$ **150,000**

18. $21,000 \div 7 =$ **3,000**

19. $40 \times 7,000 =$ **280,000**

20. $800,000 \div 400 =$ **2,000**

21. $20 \times 40,000 =$ **800,000**

22. $3,600 \div 90 =$ **40**

13

(4.NBT.A.2)

Exploring Place Value

> Numbers can be written in three ways.
> **Standard form** is a way to write a number that shows only the digits:
> 2,243
> **Expanded form** is a way to write a number that shows the place value of each digit:
> $2,000 + 200 + 40 + 3$
> **Word form** is a way to write a number using number words:
> two thousand, two hundred forty-three

Write each number in standard form.

1. $500 + 20 + 4 =$ **524**

2. $3,000 + 700 + 80 + 1 =$ **3,781**

3. $60,000 + 1,000 + 900 + 30 + 2 =$ **61,932**

4. $800,000 + 90,000 + 5,000 + 400 + 10 + 6 =$ **895,416**

5. $400,000 + 70,000 + 3,000 + 200 + 60 =$ **473,260**

6. six hundred thirty-one = **631**

7. seven thousand four hundred twenty-five = **7,425**

8. ninety-three thousand eight hundred seventeen = **93,817**

9. one hundred twenty-one thousand three hundred seventy-six = **121,376**

10. forty-eight thousand one hundred sixty-nine = **48,169**

14

(4.NBT.A.2)

Exploring Place Value

> Remember: **Standard form:**   35,894
> **Expanded form:** $30,000 + 5,000 + 800 + 90 + 4$
> **Word form:**   thirty-five thousand eight hundred ninety-four

Write each number in standard form.

1. $70,000 + 1,000 + 600 + 90 + 2 =$ **71,692**

2. $800,000 + 60,000 + 900 + 30 + 7 =$ **860,937**

3. twenty-four thousand eight hundred thirty-six = **24,836**

Write each number in expanded form.

4. $2,891 =$ **2,000 + 800 + 90 + 1**

5. $195,720 =$ **100,000 + 90,000 + 5,000 + 700 + 20**

6. nine thousand four hundred seventy-five = **9,000 + 400 + 70 + 5**

7. three hundred seventy thousand five hundred forty-one =
300,000 + 70,000 + 500 + 40 + 1

Write each number in word form.

8. $991,347 =$ **nine hundred ninety-one thousand three hundred forty-seven**

9. $30,724 =$ **thirty thousand seven hundred twenty-four**

10. $20,000 + 4,000 + 700 + 30 + 1 =$
twenty-four thousand seven hundred thirty-one

15

(4.NBT.A.2)

Exploring Place Value

Write each number in standard form, expanded form, or word form.

1. $40 + 3,000 + 1 + 800 + 70,000$ in standard form is **73,841**

2. $60,000 + 2,000 + 50 + 2$ in word form is **sixty-two thousand fifty-two**

3. $904,023$ in expanded form is **900,000 + 4,000 + 20 + 3**

4. $874,450$ in expanded form is **800,000 + 70,000 + 4,000 + 400 + 50**

5. five hundred two thousand eleven in standard form is **502,011**

6. $100,034$ in word form is **one hundred thousand thirty-four**

7. $600 + 700,000 + 9,000$ in word form is **seven hundred nine thousand six hundred**

8. twelve thousand twelve in expanded form is **10,000 + 2,000 + 10 + 2**

9. six hundred forty thousand two hundred nine in standard form is **640,209**

10. one million in standard form is **1,000,000**

11. $801,060$ in expanded form is **800,000 + 1,000 + 60**

12. twenty-six thousand five hundred ten in expanded form is **20,000 + 6,000 + 500 + 10**

13. $1 + 10,000 + 10 + 100,000$ in word form is **one hundred ten thousand eleven**

14. $202,202$ in word form is **two hundred two thousand two hundred two**

15. $671,000$ in expanded form is **600,000 + 70,000 + 1,000**

16. $500,000 + 7$ in word form is **five hundred thousand seven**

17. thirty-three thousand three in standard form is **33,003**

16

Answer Key

Worksheet 1 (page 17)

Name _____ 4.NBT.A.2

Comparing Numbers

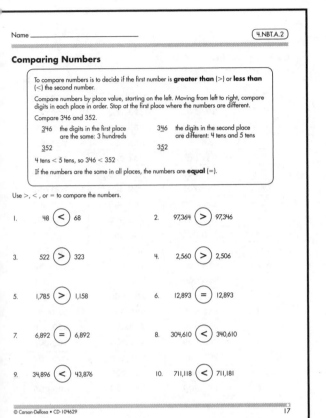

To compare numbers is to decide if the first number is **greater than** (>) or **less than** (<) the second number.

Compare numbers by place value, starting on the left. Moving from left to right, compare digits in each place in order. Stop at the first place where the numbers are different.

Compare 346 and 352.

3<u>4</u>6   the digits in the first place are the same: 3 hundreds

3<u>4</u>6   the digits in the second place are different: 4 tens and 5 tens

352

352

4 tens < 5 tens, so 346 < 352

If the numbers are the same in all places, the numbers are **equal** (=).

Use >, <, or = to compare the numbers.

1. 48 (<) 68   2. 97,364 (>) 97,346

3. 522 (>) 323   4. 2,560 (>) 2,506

5. 1,785 (>) 1,158   6. 12,893 (=) 12,893

7. 6,892 (=) 6,892   8. 304,610 (<) 340,610

9. 34,896 (<) 43,876   10. 711,118 (<) 711,181

© Carson-Dellosa • CD-104629   17

Worksheet 2 (page 18)

Name _____ 4.NBT.A.2

Comparing Numbers

Remember: Compare numbers by looking at each place value from left to right.

Use >, <, or = to compare the numbers.

1. 51 (>) 31   2. 419 (>) 411

3. 602 (>) 206   4. 35,267 (=) 35,267

5. 2,470 (>) 2,047   6. 72,380 (<) 72,387

7. 760,355 (>) 50,366   8. 21,360 (>) 21,306

9. 6,642 (<) 66,403   10. 631,207 (>) 62,746

11. 300,007 (<) 300,008   12. 9,731 (>) 973

13. 4,388 (=) 4,388   14. 7,499 (<) 7,500

15. 85,104 (<) 95,104   16. 1,877 (>) 1,766

17. 1,347 (<) 1,374   18. 204,963 (>) 201,652

18   © Carson-Dellosa • CD-104629

Worksheet 3 (page 19)

Name _____ 4.NBT.A.2

Comparing Numbers

Use >, <, or = to compare the numbers.

1. 28,941 (>) 28,914   2. 38,205 (>) 3,205   3. 620,488 (<) 628,048

4. 130,003 (<) 130,030   5. 900,267 (=) 900,267   6. 100,480 (<) 1,000,000

7. 57,352 (<) 75,352   8. 741,970 (>) 741,790   9. 412,214 (<) 412,412

10. 891,745 (>) 891,145   11. 16,661 (<) 106,661   12. 88,083 (>) 83,083

13. 40,000 + 2,000 + 700 + 1 (=) 40,000 + 2,000 + 700 + 1

14. 600,000 + 80,000 + 400 + 30 (<) 60,000 + 800,000 + 400 + 30

15. 7,000 + 500,000 + 100 + 20 + 9 (>) 20 + 7,000 + 100,000 + 500 + 9

16. 300,000 + 10,000 + 6,000 + 90 (>) 90 + 6,000 + 100 + 300,000

17. 2 + 50 + 800 + 7,000 + 90,000 + 400,000 (=) 400,000 + 800 + 50 + 90,000 + 2 + 7,000

18. three hundred thousand forty-five (<) three hundred forty-five thousand

19. twenty-eight thousand six hundred one (>) twenty-eight thousand one hundred six

20. nine hundred thirteen thousand three hundred nineteen (<) nine hundred thirty thousand three hundred nine

© Carson-Dellosa • CD-104629   19

Worksheet 4 (page 20)

Name _____ 4.NBT.A.3

Rounding Numbers

Rounding numbers is a way of replacing one number with another number that tells about how many or how much.

When rounding to a place value, look at the place value right before it. If it is 0, 1, 2, 3, or 4, round down. If it is 5, 6, 7, 8, or 9, round up.

Round 23 to the nearest ten.
Look at the ones digit.

20   2<u>3</u>   30

Round 23 down to 20.

Round 284 to the nearest hundred.
Look at the tens digit.

200   2<u>8</u>4   300

Round 284 up to 300.

Round the numbers in each set to the place value listed.

1. ten
24 __**20**__   567 __**570**__   7,075 __**7,080**__

2. hundred
26,483 __**26,500**__   809 __**800**__   4,095 __**4,100**__

3. thousand
873,609 __**874,000**__   55,937 __**56,000**__   2,148 __**2,000**__

4. ten thousand
902,813 __**900,000**__   66,408 __**70,000**__   40,742 __**40,000**__

5. hundred thousand
720,311 __**700,000**__   485,407 __**500,000**__   673,054 __**700,000**__

20   © Carson-Dellosa • CD-104629

Answer Key

Name _____   4.NBT.A.3

Rounding Numbers

> Remember: If the digit to the right is less than 5, round down.
> If the digit to the right is 5 or greater, round up.

Round the numbers to the underlined place value.

1. 5_8_2 **580**
2. 6_5_,914 **70,000**
3. 5_7_2,133 **570,000**

4. 3_1_9 **320**
5. 87,4_0_8 **87,410**
6. 60,8_4_7 **60,850**

7. _7_55 **800**
8. 3_2_5,035 **330,000**
9. _4_92,795 **500,000**

10. 9,3_2_1 **9,320**
11. 98_2_,485 **982,000**
12. 381,_2_91 **381,300**

13. 8,0_9_2 **8,100**
14. _6_45,062 **600,000**
15. 59,9_9_9 **60,000**

16. _4_,902 **5,000**
17. 768,8_9_1 **768,900**
18. 10_9_,900 **110,000**

19. 8_2_,730 **83,000**
20. 20_4_,519 **205,000**
21. _9_78,010 **1,000,000**

Name _____   4.NBT.A.3

Rounding Numbers

Round the numbers to each place value.

		hundred	thousand	hundred thousand
1.	128,549	128,500	129,000	100,000
2.	395,305	395,300	395,000	400,000
3.	802,381	802,400	802,000	800,000
4.	654,728	654,700	655,000	700,000
5.	930,960	931,000	931,000	900,000
6.	419,283	419,300	419,000	400,000
7.	728,675	728,700	729,000	700,000
8.	592,014	592,000	592,000	600,000
9.	265,529	265,500	266,000	300,000
10.	139,993	140,000	140,000	100,000
11.	539,102	539,100	539,000	500,000
12.	984,241	984,200	984,000	1,000,000
13.	426,009	426,000	426,000	400,000
14.	737,092	737,100	737,000	700,000
15.	643,598	643,600	644,000	600,000
16.	849,251	849,300	849,000	800,000
17.	489,543	489,500	490,000	500,000
18.	265,877	265,900	266,000	300,000
19.	100,035	100,000	100,000	100,000
20.	574,892	574,900	575,000	600,000

Name _____   4.NBT.B.4

Adding and Subtracting Large Numbers

> Sometimes it is necessary to regroup when adding or subtracting large numbers.
>
> ```
>  1
>  53      Regroup the number 12 into
> + 19     1 ten and 2 ones.
> ----     Carry the 1 ten to the tens column.
>  72      Finish by adding the tens.
> ```
>
> ```
>  4 13
>  5̸3̸      6 is too big to subtract from 3.
> - 16     Borrow a ten and regroup it into
> ----     10 ones.
>  37      Subtract the ones column.
>          Subtract the tens column.
> ```

Solve each problem. Regroup when necessary.

1. 6,376 + 2,019 = **8,395**
2. 2,393 + 4,392 = **6,785**
3. 8,293 + 4,239 = **12,532**
4. 3,768 + 5,949 = **9,717**
5. 1,665 + 3,773 = **5,438**
6. 2,343 + 7,328 = **9,671**

7. 7,320 + 5,394 = **12,714**
8. 9,347 + 7,323 = **16,670**
9. 8,659 + 9,347 = **18,006**
10. 3,424 + 9,483 = **12,907**
11. 6,784 + 1,296 = **8,080**
12. 4,392 + 4,959 = **9,351**

13. 9,534 − 2,389 = **7,145**
14. 5,464 − 2,756 = **2,708**
15. 3,526 − 1,653 = **1,873**
16. 3,354 − 2,328 = **1,026**
17. 5,247 − 3,836 = **1,411**
18. 8,456 − 3,462 = **4,994**

19. 4,755 − 3,875 = **880**
20. 7,243 − 2,376 = **4,867**
21. 6,845 − 4,764 = **2,081**
22. 5,935 − 3,837 = **2,098**
23. 4,376 − 2,438 = **1,938**
24. 9,122 − 4,547 = **4,575**

Name _____   4.NBT.B.4

Adding and Subtracting Large Numbers

> When subtracting, you may need to borrow from a zero.
> To do this, borrow to make the zero a ten. Then, borrow from the ten.
>
> ```
>    9
>  4 1̸0 12
>  5̸0̸2̸     Borrow from the 5 hundred to make 10 tens.
> - 346    Then, borrow a ten to make 12 ones.
> -----
>  156
> ```

Solve each problem. Regroup when necessary.

1. 84,936 + 25,432 = **110,368**
2. 79,675 + 14,283 = **93,958**
3. 35,349 + 36,393 = **71,742**
4. 26,434 + 16,398 = **42,832**
5. 49,231 + 15,332 = **64,563**
6. 37,221 + 22,418 = **59,639**

7. 76,376 + 52,019 = **128,395**
8. 82,393 + 74,392 = **156,785**
9. 58,293 + 34,239 = **92,532**
10. 43,768 + 15,949 = **59,717**
11. 91,665 + 13,773 = **105,438**
12. 22,343 + 27,328 = **49,671**

13. 57,320 + 65,394 = **122,714**
14. 49,347 + 77,323 = **126,670**
15. 28,659 + 19,347 = **48,006**
16. 43,768 + 15,949 = **59,717**
17. 56,784 + 61,296 = **118,080**
18. 74,392 + 44,959 = **119,351**

19. 89,534 − 12,389 = **77,145**
20. 75,464 − 22,756 = **52,708**
21. 63,526 − 51,653 = **11,873**
22. 93,354 − 42,328 = **51,026**
23. 45,247 − 33,836 = **11,411**
24. 28,456 − 13,462 = **14,994**

25. 54,755 − 23,875 = **30,880**
26. 77,243 − 52,376 = **24,867**
27. 16,845 − 14,764 = **2,081**
28. 65,935 − 23,837 = **42,098**
29. 84,376 − 12,438 = **71,938**
30. 89,122 − 64,547 = **24,575**

31. 32,643 − 11,439 = **21,204**
32. 53,765 − 23,498 = **30,267**
33. 67,236 − 12,276 = **54,960**
34. 87,340 − 55,364 = **31,976**
35. 96,849 − 74,114 = **22,735**
36. 67,414 − 42,838 = **24,576**

Answer Key

Adding and Subtracting Large Numbers

Fill in the missing numbers to complete each problem.

1.
```
  4 6, 3 7 [2]
+ [1] 4, 6 7 4
  6 1, [0] 4 6
```

2.
```
  1 [3], 0 8 9
+ 3 7, 4 [7] 5
  5 0, 5 6 [4]
```

3.
```
  [2] 6, 1 7 8
+ 6 4, 4 [8] 8
  9 0, [6] 6 6
```

4.
```
  3 5, 4 3 [9]
+ 1 [6], 7 8 0
  5 2, 2 [1] 9
```

5.
```
  9 [2], 3 [4] 7
+ 1 9, 8 0 7
  1 1 2, 1 5 4
```

6.
```
  3 1 2, 8 5 [6]
+ 3 [7] 3, 5 8 7
  6 8 6, 4 [4] 3
```

7.
```
  9 2, 8 5 [1]
- [3] 5, 2 7 9
  5 7, [5] 7 2
```

8.
```
  [7] 3, 1 2 8
- 6 4, [3] 5 6
    8, 7 7 [2]
```

9.
```
  1 2 [8], 5 [4] 0
-   5 6, 2 0 [1]
    7 2, 3 3 9
```

10.
```
  3 9 4, 2 7 1
- [3] 0 2, 0 5 6
  9 [2] 2, 1 5
```

11.
```
  7 2, [1] 3 3
- 5 4, 8 [2] 5
  [1] 7, 3 0 8
```

12.
```
  8 [4] 9, 0 2 2
- 4 6 7, 1 3 [9]
  3 8 [1], 8 8 3
```

Multiplying Multi-Digit Numbers by One-Digit Numbers

> Sometimes it is necessary to regroup when multiplying multi-digit numbers by one-digit numbers. Regroup by carrying tens.
>
> ```
>   2
>  28
>  ×3
>  84
> ```
> $8 \times 3 = 24$
> Regroup the number 24 into 2 tens and 4 ones.
> Carry the 2 tens to the tens column.
> Multiply the tens column. $(2 \times 3 = 6)$
> Then, add the carried tens. $(6 + 2 = 8)$

Solve each problem. Regroup when necessary.

1. 45 ×2	2. 56 ×4	3. 34 ×3	4. 57 ×4	5. 28 ×3	6. 46 ×6
90	**224**	**102**	**228**	**84**	**276**

7. 39 ×6	8. 19 ×8	9. 36 ×6	10. 76 ×5	11. 44 ×5	12. 75 ×2
234	**152**	**216**	**380**	**220**	**150**

13. 27 ×6	14. 22 ×9	15. 83 ×6	16. 87 ×2	17. 37 ×3	18. 49 ×7
162	**198**	**498**	**174**	**111**	**343**

19. 74 ×3	20. 37 ×6	21. 53 ×5	22. 68 ×5	23. 38 ×4	24. 77 ×8
222	**222**	**265**	**340**	**152**	**616**

Multiplying Multi-Digit Numbers by One-Digit Numbers

Solve each problem. Regroup when necessary.

1. 323 ×5	2. 515 ×4	3. 255 ×4	4. 915 ×2	5. 860 ×2	6. 561 ×9
1,615	**2,060**	**1,020**	**1,830**	**1,720**	**5,049**

7. 109 ×4	8. 812 ×8	9. 503 ×3	10. 827 ×3	11. 122 ×8	12. 523 ×6
436	**6,496**	**1,509**	**2,481**	**976**	**3,138**

13. 206 ×5	14. 617 ×7	15. 134 ×6	16. 4,364 ×2	17. 8,436 ×5	18. 5,691 ×5
1,030	**4,319**	**804**	**8,728**	**42,180**	**28,455**

19. 1,029 ×5	20. 5,414 ×2	21. 6,501 ×7	22. 2,897 ×4	23. 7,152 ×4	24. 4,646 ×9
5,145	**10,828**	**45,507**	**11,588**	**28,608**	**41,814**

25. 5,678 ×2	26. 4,610 ×5	27. 5,129 ×5	28. 3,162 ×4	29. 7,109 ×6	30. 4,862 ×7
11,356	**23,050**	**25,645**	**12,648**	**42,654**	**34,034**

Multiplying Multi-Digit Numbers by One-Digit Numbers

Solve each problem. Regroup when necessary.

1. 4,113 ×6	2. 7,312 ×7	3. 8,900 ×8	4. 5,308 ×4	5. 4,930 ×4	6. 6,342 ×5
24,678	**51,184**	**71,200**	**21,232**	**19,720**	**31,710**

7. 4,213 ×6	8. 9,980 ×5	9. 2,794 ×7	10. 9,755 ×8	11. 3,214 ×7	12. 2,317 ×3
25,278	**49,900**	**19,558**	**78,040**	**22,498**	**6,951**

13. 6,746 ×3	14. 6,677 ×4	15. 8,227 ×2	16. 5,857 ×3	17. 3,351 ×5	18. 2,356 ×3
20,238	**26,708**	**16,454**	**17,571**	**16,755**	**7,068**

19. 4,845 ×2	20. 7,934 ×3	21. 4,065 ×6	22. 2,132 ×6	23. 7,021 ×4	24. 9,442 ×3
9,690	**23,802**	**24,390**	**12,792**	**28,084**	**28,326**

25. 4,365 ×6	26. 3,225 ×5	27. 8,222 ×4	28. 7,422 ×5	29. 8,265 ×3	30. 7,120 ×2
26,190	**16,125**	**32,888**	**37,110**	**24,795**	**14,240**

31. 7,322 ×6	32. 6,434 ×6	33. 7,387 ×5	34. 8,483 ×4	35. 8,612 ×6	36. 6,947 ×5
43,932	**38,604**	**36,935**	**33,932**	**51,672**	**34,735**

Answer Key

Answer Key

Name _____ 4.NBT.B.6

Dividing without Remainders

> Sometimes the first digit in the dividend is not large enough to divide into. Move to the next digit in the dividend. Divide into that two-digit dividend, multiply, and subtract.
>
> ```
>      34
> 6 )204
>   - 18
>     24
>   - 24
>      0
> ```

Solve each problem.

1. **83** 5)415
2. **52** 3)156
3. **22** 7)154
4. **87** 8)696

5. **153** 4)612
6. **98** 8)784
7. **81** 4)324
8. **174** 4)696

9. **71** 6)426
10. **53** 3)159
11. **89** 4)356
12. **95** 5)475

13. **66** 2)132
14. **59** 6)354
15. **62** 6)372
16. **74** 2)148

17. **87** 4)348
18. **73** 2)146
19. **63** 5)315
20. **80** 4)320

33

Name _____ 4.NBT.B.6

Dividing without Remainders

Solve each problem.

1. **152** 9)1,368
2. **307** 4)1,228
3. **674** 8)5,392
4. **313** 6)1,878

5. **279** 5)1,395
6. **418** 7)2,926
7. **252** 4)1,008
8. **195** 5)975

9. **532** 4)2,128
10. **612** 2)1,224
11. **451** 6)2,706
12. **673** 3)2,019

13. **336** 3)1,008
14. **486** 8)3,888
15. **203** 7)1,421
16. **225** 5)1,125

17. **512** 2)1,024
18. **378** 3)1,134
19. **620** 8)4,960
20. **310** 9)2,790

34

Name _____ 4.NBT.B.6

Dividing with Remainders

> Sometimes it is necessary to name a remainder when dividing. The **remainder** is the number remaining after the division is complete.
>
> 3 cannot divide into 2. There are no more digits to bring down from the dividend. The difference becomes the remainder (r).
>
> ```
>      29r2
> 3 )89
>   - 6
>    29
>  - 27
>     2
> ```

Solve each problem.

1. **11r5** 7)82
2. **13r2** 4)54
3. **8r2** 3)26

4. **11r7** 8)95
5. **4r2** 4)18
6. **8r1** 7)57

7. **15r3** 4)63
8. **4r2** 5)22
9. **3r3** 5)18

10. **16r1** 5)81
11. **10r1** 4)41
12. **9r2** 3)29

13. **12r2** 6)74
14. **4r5** 8)37
15. **8r2** 5)42

35

Name _____ 4.NBT.B.6

Dividing with Remainders

Solve each problem.

1. **218r1** 4)873
2. **188r3** 5)943
3. **119r5** 8)957
4. **109r6** 9)987

5. **130r5** 7)915
6. **105r2** 5)527
7. **298r1** 2)597
8. **108r1** 9)973

9. **143r2** 4)574
10. **108r5** 6)653
11. **261r1** 3)784
12. **121r2** 4)486

13. **209r2** 3)629
14. **150r1** 2)301
15. **127r2** 5)637
16. **215r2** 4)862

17. **366r1** 2)733
18. **117r1** 8)937
19. **191r1** 3)574
20. **163r1** 4)653

36

Answer Key

Worksheet 1 (top left)

Dividing with Remainders

Solve each problem. Show your work on another sheet of paper. Write your answers here.

1. **1,231r5** $6\overline{)7,391}$
2. **958** $3\overline{)2,874}$
3. **2,079r1** $3\overline{)6,238}$

4. **547** $8\overline{)4,376}$
5. **627r2** $6\overline{)3,764}$
6. **313r2** $9\overline{)2,819}$

7. **4,248r1** $2\overline{)8,497}$
8. **1,358r1** $6\overline{)8,149}$
9. **676r1** $5\overline{)3,381}$

10. **746r3** $4\overline{)2,987}$
11. **1,148r4** $7\overline{)8,040}$
12. **1,262r2** $3\overline{)3,788}$

13. **714r3** $7\overline{)5,001}$
14. **3,420 r1** $2\overline{)6,841}$
15. **1,578r1** $6\overline{)9,469}$

16. **1,065r3** $5\overline{)5,328}$
17. **1,476r4** $5\overline{)7,384}$
18. **1,494r2** $4\overline{)5,978}$

19. **384r2** $4\overline{)1,538}$
20. **2,405r1** $2\overline{)4,811}$
21. **1,228r2** $7\overline{)8,598}$

22. **2,143** $4\overline{)8,572}$
23. **2,314r1** $3\overline{)6,943}$
24. **1,072r4** $6\overline{)6,436}$

25. **585r7** $8\overline{)4,687}$
26. **1,047r2** $5\overline{)5,237}$
27. **684r2** $7\overline{)4,790}$

28. **697r1** $5\overline{)3,486}$
29. **2,258r3** $4\overline{)9,035}$
30. **571r4** $7\overline{)4,001}$

37

Worksheet 2 (top right)

Finding Equivalent Fractions

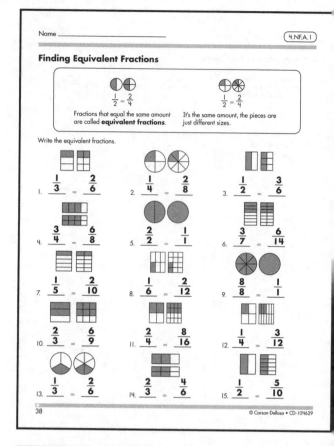

Fractions that equal the same amount are called **equivalent fractions**.   It's the same amount, the pieces are just different sizes.

Write the equivalent fractions.

1. $\dfrac{1}{3} = \dfrac{2}{6}$
2. $\dfrac{1}{4} = \dfrac{2}{8}$
3. $\dfrac{1}{2} = \dfrac{3}{6}$

4. $\dfrac{3}{4} = \dfrac{6}{8}$
5. $\dfrac{2}{2} = \dfrac{1}{1}$
6. $\dfrac{3}{7} = \dfrac{6}{14}$

7. $\dfrac{1}{5} = \dfrac{2}{10}$
8. $\dfrac{1}{6} = \dfrac{2}{12}$
9. $\dfrac{8}{8} = \dfrac{1}{1}$

10. $\dfrac{2}{3} = \dfrac{6}{9}$
11. $\dfrac{2}{4} = \dfrac{8}{16}$
12. $\dfrac{1}{4} = \dfrac{3}{12}$

13. $\dfrac{1}{3} = \dfrac{2}{6}$
14. $\dfrac{2}{3} = \dfrac{4}{6}$
15. $\dfrac{1}{2} = \dfrac{5}{10}$

38

Worksheet 3 (bottom left)

Finding Equivalent Fractions

To find equivalent fractions, multiply the numerator and denominator by the same number.

Write the missing numerators to make the fractions in each row equivalent. Draw one pair of equivalent fractions for each problem.

1. $\dfrac{1}{2} = \dfrac{18}{36} = \dfrac{9}{18} = \dfrac{8}{16} = \dfrac{21}{42} = \dfrac{24}{48}$

Drawings will vary.

2. $\dfrac{5}{6} = \dfrac{40}{48} = \dfrac{10}{12} = \dfrac{25}{30} = \dfrac{15}{18} = \dfrac{20}{24}$

3. $\dfrac{1}{3} = \dfrac{3}{9} = \dfrac{9}{27} = \dfrac{30}{90} = \dfrac{2}{6} = \dfrac{4}{12}$

4. $\dfrac{3}{4} = \dfrac{18}{24} = \dfrac{12}{16} = \dfrac{6}{8} = \dfrac{15}{20} = \dfrac{27}{36}$

5. $\dfrac{4}{9} = \dfrac{8}{18} = \dfrac{20}{45} = \dfrac{16}{36} = \dfrac{24}{54} = \dfrac{12}{27}$

6. $\dfrac{7}{8} = \dfrac{14}{16} = \dfrac{49}{56} = \dfrac{21}{24} = \dfrac{42}{48} = \dfrac{28}{32}$

7. $\dfrac{3}{7} = \dfrac{9}{21} = \dfrac{18}{42} = \dfrac{6}{14} = \dfrac{15}{35} = \dfrac{12}{28}$

8. $\dfrac{2}{5} = \dfrac{20}{50} = \dfrac{4}{10} = \dfrac{16}{40} = \dfrac{6}{15} = \dfrac{10}{25}$

39

Worksheet 4 (bottom right)

Finding Equivalent Fractions

Write the missing numerator to make each pair equivalent.

1. $\dfrac{2}{3} = \dfrac{8}{12}$
2. $\dfrac{8}{9} = \dfrac{48}{54}$
3. $\dfrac{1}{2} = \dfrac{5}{10}$

4. $\dfrac{1}{8} = \dfrac{4}{32}$
5. $\dfrac{4}{9} = \dfrac{36}{81}$
6. $\dfrac{2}{9} = \dfrac{4}{18}$

7. $\dfrac{3}{4} = \dfrac{12}{16}$
8. $\dfrac{1}{2} = \dfrac{6}{12}$
9. $\dfrac{4}{5} = \dfrac{20}{25}$

10. $\dfrac{2}{5} = \dfrac{12}{30}$
11. $\dfrac{7}{8} = \dfrac{56}{64}$
12. $\dfrac{2}{3} = \dfrac{10}{15}$

13. $\dfrac{2}{5} = \dfrac{4}{10}$
14. $\dfrac{3}{8} = \dfrac{6}{16}$
15. $\dfrac{5}{8} = \dfrac{15}{24}$

16. $\dfrac{3}{4} = \dfrac{18}{24}$
17. $\dfrac{3}{5} = \dfrac{9}{15}$
18. $\dfrac{3}{7} = \dfrac{6}{14}$

19. $\dfrac{1}{6} = \dfrac{2}{12}$
20. $\dfrac{4}{5} = \dfrac{16}{20}$
21. $\dfrac{3}{7} = \dfrac{9}{21}$

22. $\dfrac{5}{6} = \dfrac{35}{42}$
23. $\dfrac{1}{6} = \dfrac{6}{36}$
24. $\dfrac{5}{8} = \dfrac{25}{40}$

40

Answer Key

Name _____

(4.NF.A.1)

Reducing Fractions

To put a fraction in **simplest form** means to rename or **reduce** the fraction without changing the amount.

Follow the steps to reduce a fraction.

$\frac{6}{9}$ 1. Find the largest number that can be divided into the numerator and denominator.

$\div 3$ 2. Both numbers can be divided by 3. $\frac{6}{9} \div \frac{3}{3} = \frac{2}{3}$

$\frac{2}{3}$ 3. The simplest form of $\frac{6}{9}$ is $\frac{2}{3}$.

The two fractions still represent the same amount.

Reduce each fraction to simplest form.

1. $\frac{4}{10} \div \frac{2}{2} = \frac{2}{5}$ 2. $\frac{8}{32} \div \frac{8}{8} = \frac{1}{4}$ 3. $\frac{10}{12} \div \frac{2}{2} = \frac{5}{6}$

4. $\frac{18}{27} \div \frac{9}{9} = \frac{2}{3}$ 5. $\frac{5}{15} \div \frac{5}{5} = \frac{1}{3}$ 6. $\frac{4}{26} \div \frac{2}{2} = \frac{2}{13}$

7. $\frac{16}{56} \div \frac{8}{8} = \frac{2}{7}$ 8. $\frac{20}{45} \div \frac{5}{5} = \frac{4}{9}$ 9. $\frac{18}{40} \div \frac{2}{2} = \frac{9}{20}$

Name _____

(4.NF.A.1)

Reducing Fractions

Write each fraction in simplest form.

1. $\frac{6}{8} = \frac{3}{4}$ 2. $\frac{3}{24} = \frac{1}{8}$ 3. $\frac{20}{35} = \frac{4}{7}$

4. $\frac{15}{20} = \frac{3}{4}$ 5. $\frac{10}{20} = \frac{1}{2}$ 6. $\frac{6}{16} = \frac{3}{8}$

7. $\frac{5}{20} = \frac{1}{4}$ 8. $\frac{4}{8} = \frac{1}{2}$ 9. $\frac{4}{16} = \frac{1}{4}$

10. $\frac{6}{9} = \frac{2}{3}$ 11. $\frac{4}{20} = \frac{1}{5}$ 12. $\frac{3}{15} = \frac{1}{5}$

13. $\frac{3}{12} = \frac{1}{4}$ 14. $\frac{5}{15} = \frac{1}{3}$ 15. $\frac{8}{16} = \frac{1}{2}$

16. $\frac{7}{21} = \frac{1}{3}$ 17. $\frac{5}{25} = \frac{1}{5}$ 18. $\frac{15}{30} = \frac{1}{2}$

19. $\frac{2}{8} = \frac{1}{4}$ 20. $\frac{14}{21} = \frac{2}{3}$ 21. $\frac{12}{16} = \frac{3}{4}$

Name _____

(4.NF.A.1)

Reducing Fractions

Write each fraction in simplest form.

1. $\frac{4}{8} = \frac{1}{2}$ 2. $\frac{7}{14} = \frac{1}{2}$ 3. $\frac{20}{30} = \frac{2}{3}$

4. $\frac{10}{28} = \frac{5}{14}$ 5. $\frac{14}{40} = \frac{7}{20}$ 6. $\frac{6}{20} = \frac{3}{10}$

7. $\frac{4}{12} = \frac{1}{3}$ 8. $\frac{2}{8} = \frac{1}{4}$ 9. $\frac{5}{30} = \frac{1}{6}$

10. $\frac{3}{9} = \frac{1}{3}$ 11. $\frac{2}{6} = \frac{1}{3}$ 12. $\frac{3}{15} = \frac{1}{5}$

13. $\frac{3}{12} = \frac{1}{4}$ 14. $\frac{8}{24} = \frac{1}{3}$ 15. $\frac{8}{20} = \frac{2}{5}$

16. $\frac{6}{18} = \frac{1}{3}$ 17. $\frac{5}{20} = \frac{1}{4}$ 18. $\frac{15}{20} = \frac{3}{4}$

19. $\frac{2}{4} = \frac{1}{2}$ 20. $\frac{15}{21} = \frac{5}{7}$ 21. $\frac{12}{30} = \frac{2}{5}$

22. $\frac{20}{22} = \frac{10}{11}$ 23. $\frac{7}{28} = \frac{1}{4}$ 24. $\frac{16}{32} = \frac{1}{2}$

25. $\frac{12}{15} = \frac{4}{5}$ 26. $\frac{18}{24} = \frac{3}{4}$ 27. $\frac{4}{18} = \frac{2}{9}$

28. $\frac{5}{15} = \frac{1}{3}$ 29. $\frac{15}{20} = \frac{3}{4}$ 30. $\frac{21}{45} = \frac{7}{15}$

Name _____

(4.NF.A.1)

Writing Improper Fractions as Mixed Numbers

When the numerator is greater than or equal to the denominator, it is called an **improper fraction**.

$\frac{9}{4}$

When a whole number is with a fraction, it is called a **mixed number**. An improper fraction ($\frac{9}{4}$) can be changed to a mixed number.

$2\frac{1}{4}$

Divide the numerator by the denominator.

The quotient becomes the whole number. The remainder becomes a fraction. Use the denominator of the improper fraction.

$\frac{9}{4}$ $4\overline{)9}^{\,2r1}$ -8 $\overline{1}$

$2\frac{1}{4}$

Write each improper fraction as a mixed number.

1. $\frac{14}{3} = 4\frac{2}{3}$ 2. $\frac{22}{7} = 3\frac{1}{7}$ 3. $\frac{44}{8} = 5\frac{4}{8}$ 4. $\frac{32}{5} = 6\frac{2}{5}$

5. $\frac{13}{4} = 3\frac{1}{4}$ 6. $\frac{40}{6} = 6\frac{4}{6}$ 7. $\frac{18}{4} = 4\frac{2}{4}$ 8. $\frac{59}{9} = 6\frac{5}{9}$

9. $\frac{15}{8} = 1\frac{7}{8}$ 10. $\frac{23}{9} = 2\frac{5}{9}$ 11. $\frac{32}{7} = 4\frac{4}{7}$ 12. $\frac{44}{6} = 7\frac{2}{6}$

13. $\frac{12}{5} = 2\frac{2}{5}$ 14. $\frac{13}{5} = 2\frac{3}{5}$ 15. $\frac{58}{9} = 6\frac{4}{9}$ 16. $\frac{15}{4} = 3\frac{3}{4}$

Answer Key

Name _____

(4.NF.A.1)

Writing Improper Fractions as Mixed Numbers
Write each improper fraction as a mixed number in simplest form.

1. $\frac{4}{3} = 1\frac{1}{3}$
2. $\frac{20}{15} = 1\frac{1}{3}$
3. $\frac{7}{4} = 1\frac{3}{4}$

4. $\frac{55}{12} = 4\frac{7}{12}$
5. $\frac{18}{5} = 3\frac{3}{5}$
6. $\frac{5}{2} = 2\frac{1}{2}$

7. $\frac{5}{3} = 1\frac{2}{3}$
8. $\frac{12}{5} = 2\frac{2}{5}$
9. $\frac{13}{4} = 3\frac{1}{4}$

10. $\frac{15}{6} = 2\frac{1}{2}$
11. $\frac{13}{2} = 6\frac{1}{2}$
12. $\frac{17}{9} = 1\frac{8}{9}$

13. $\frac{10}{4} = 2\frac{1}{2}$
14. $\frac{19}{2} = 9\frac{1}{2}$
15. $\frac{27}{5} = 5\frac{2}{5}$

16. $\frac{15}{4} = 3\frac{3}{4}$
17. $\frac{8}{3} = 2\frac{2}{3}$
18. $\frac{15}{8} = 1\frac{7}{8}$

19. $\frac{6}{4} = 1\frac{1}{2}$
20. $\frac{43}{7} = 6\frac{1}{7}$
21. $\frac{19}{11} = 1\frac{8}{11}$

22. $\frac{20}{7} = 2\frac{6}{7}$
23. $\frac{9}{4} = 2\frac{1}{4}$
24. $\frac{17}{4} = 4\frac{1}{4}$

Name _____

(4.NF.A.1)

Writing Improper Fractions as Mixed Numbers
Write each improper fraction as a mixed number in simplest form.

1. $\frac{6}{4} = 1\frac{1}{2}$
2. $\frac{21}{12} = 1\frac{3}{4}$
3. $\frac{9}{4} = 2\frac{1}{4}$

4. $\frac{25}{11} = 2\frac{3}{11}$
5. $\frac{19}{5} = 3\frac{4}{5}$
6. $\frac{3}{2} = 1\frac{1}{2}$

7. $\frac{7}{4} = 1\frac{3}{4}$
8. $\frac{13}{3} = 4\frac{1}{3}$
9. $\frac{14}{6} = 2\frac{1}{3}$

10. $\frac{16}{5} = 3\frac{1}{5}$
11. $\frac{13}{5} = 2\frac{3}{5}$
12. $\frac{14}{8} = 1\frac{3}{4}$

13. $\frac{11}{2} = 5\frac{1}{2}$
14. $\frac{17}{4} = 4\frac{1}{4}$
15. $\frac{19}{2} = 9\frac{1}{2}$

16. $\frac{25}{3} = 8\frac{1}{3}$
17. $\frac{8}{3} = 2\frac{2}{3}$
18. $\frac{11}{6} = 1\frac{5}{6}$

19. $\frac{10}{3} = 3\frac{1}{3}$
20. $\frac{33}{6} = 5\frac{1}{2}$
21. $\frac{14}{9} = 1\frac{5}{9}$

22. $\frac{20}{8} = 2\frac{1}{2}$
23. $\frac{7}{4} = 1\frac{3}{4}$
24. $\frac{13}{3} = 4\frac{1}{3}$

25. $\frac{12}{5} = 2\frac{2}{5}$
26. $\frac{18}{11} = 1\frac{7}{11}$
27. $\frac{9}{2} = 4\frac{1}{2}$

28. $\frac{15}{4} = 3\frac{3}{4}$
29. $\frac{10}{6} = 1\frac{2}{3}$
30. $\frac{10}{4} = 2\frac{1}{2}$

Name _____

(4.NF.A.1)

Writing Mixed Numbers as Improper Fractions

When the numerator is greater than or equal to the denominator, it is called an **improper fraction**. $\frac{9}{4}$

When a whole number is with a fraction, it is called a **mixed number**. $2\frac{1}{4}$

A mixed number ($2\frac{1}{4}$) can be changed to an improper fraction.

Multiply the whole number by the denominator.	Add the product to the numerator.	Write the sum over the original denominator.
$2\frac{1}{4}$ $2 \times 4 = 8$	$8 + 1 = 9$	$\frac{9}{4}$

Write each mixed number as an improper fraction.

1. $1\frac{4}{3} = \frac{7}{3}$
2. $3\frac{1}{7} = \frac{22}{7}$
3. $4\frac{4}{8} = \frac{36}{8}$
4. $6\frac{2}{5} = \frac{32}{5}$

5. $3\frac{3}{4} = \frac{15}{4}$
6. $7\frac{4}{6} = \frac{46}{6}$
7. $5\frac{1}{4} = \frac{21}{4}$
8. $6\frac{5}{9} = \frac{59}{9}$

9. $1\frac{7}{8} = \frac{15}{8}$
10. $2\frac{5}{9} = \frac{23}{9}$
11. $4\frac{4}{7} = \frac{32}{7}$
12. $7\frac{2}{6} = \frac{44}{6}$

13. $2\frac{2}{5} = \frac{12}{5}$
14. $2\frac{3}{5} = \frac{13}{5}$
15. $6\frac{4}{9} = \frac{58}{9}$
16. $3\frac{3}{4} = \frac{15}{4}$

Name _____

(4.NF.A.1)

Writing Mixed Numbers as Improper Fractions
Write each mixed number as an improper fraction.

1. $3\frac{1}{2} = \frac{7}{2}$
2. $5\frac{7}{8} = \frac{47}{8}$
3. $7\frac{4}{5} = \frac{39}{5}$

4. $1\frac{1}{10} = \frac{11}{10}$
5. $6\frac{5}{8} = \frac{53}{8}$
6. $5\frac{2}{3} = \frac{17}{3}$

7. $9\frac{1}{2} = \frac{19}{2}$
8. $4\frac{3}{8} = \frac{35}{8}$
9. $8\frac{2}{3} = \frac{26}{3}$

10. $2\frac{2}{3} = \frac{8}{3}$
11. $2\frac{4}{9} = \frac{22}{9}$
12. $4\frac{3}{4} = \frac{19}{4}$

13. $2\frac{3}{8} = \frac{19}{8}$
14. $4\frac{2}{4} = \frac{18}{4}$
15. $6\frac{5}{7} = \frac{47}{7}$

16. $10\frac{3}{5} = \frac{53}{5}$
17. $4\frac{5}{9} = \frac{41}{9}$
18. $7\frac{5}{6} = \frac{47}{6}$

Answer Key

Name _____   4.NF.A.1

Writing Mixed Numbers as Improper Fractions
Write each mixed number as an improper fraction.

1. $1\frac{2}{3} = \frac{5}{3}$
2. $10\frac{5}{6} = \frac{65}{6}$
3. $5\frac{4}{5} = \frac{29}{5}$

4. $2\frac{1}{12} = \frac{25}{12}$
5. $12\frac{4}{5} = \frac{64}{5}$
6. $1\frac{1}{5} = \frac{6}{5}$

7. $7\frac{1}{6} = \frac{43}{6}$
8. $9\frac{6}{8} = \frac{78}{8}$
9. $9\frac{2}{8} = \frac{74}{8}$

10. $20\frac{2}{8} = \frac{162}{8}$
11. $8\frac{3}{9} = \frac{75}{9}$
12. $3\frac{3}{8} = \frac{27}{8}$

13. $2\frac{3}{4} = \frac{11}{4}$
14. $1\frac{1}{4} = \frac{5}{4}$
15. $15\frac{3}{4} = \frac{63}{4}$

16. $7\frac{10}{16} = \frac{122}{16}$
17. $11\frac{3}{9} = \frac{102}{9}$
18. $12\frac{13}{15} = \frac{193}{15}$

19. $3\frac{2}{5} = \frac{17}{5}$
20. $6\frac{1}{3} = \frac{19}{3}$
21. $7\frac{2}{5} = \frac{37}{5}$

Name _____   4.NF.A.2

Comparing Fractions
Use the fraction bars to compare the fractions.

1. Circle the fraction that is less.
- $\frac{1}{2}$ (⊘ $\frac{2}{5}$)
- $\frac{2}{6}$ (⊘ $\frac{2}{8}$)
- (⊘ $\frac{4}{6}$) $\frac{3}{4}$
- $\frac{4}{10}$ (⊘ $\frac{1}{3}$)

2. Circle the fraction that is greater.
- $\frac{3}{6}$ (⊘ $\frac{3}{5}$)
- $\frac{4}{8}$ (⊘ $\frac{4}{5}$)
- (⊘ $\frac{9}{10}$) $\frac{2}{3}$
- (⊘ $\frac{4}{10}$) $\frac{2}{6}$

Name _____   4.NF.A.1, 4.NF.A.2

Comparing Fractions

In order to compare, you must find equivalent fractions. $\frac{1}{4}\bigcirc\frac{1}{8}$

First, we must find a common denominator. Does something multiplied by 4 equal 8? Yes, 2. 8 is our common denominator.

Multiply the numerator and denominator by 2 to create equivalent fractions. $\frac{1}{4}=\frac{2}{8}$
$\frac{1\times2}{4\times2}=\frac{2}{8}$
Compare. $\frac{1}{4}>\frac{1}{8}$

Identify each fraction. Use > or < to compare the fractions.

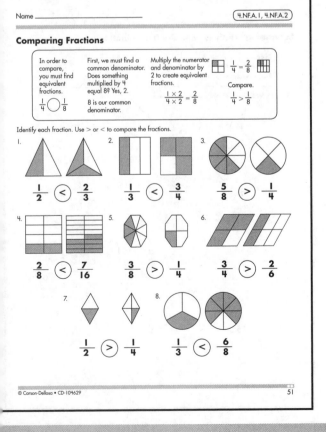

1. $\frac{1}{2} < \frac{2}{3}$
2. $\frac{1}{3} < \frac{3}{4}$
3. $\frac{5}{8} > \frac{1}{4}$
4. $\frac{2}{8} < \frac{7}{16}$
5. $\frac{3}{8} > \frac{1}{4}$
6. $\frac{3}{4} > \frac{2}{6}$
7. $\frac{1}{2} > \frac{1}{4}$
8. $\frac{1}{3} < \frac{6}{8}$

Name _____   4.NF.A.1, 4.NF.A.2

Comparing Fractions
Use >, <, or = to compare the fractions.

1. $\frac{5}{10} > \frac{2}{10}$
2. $\frac{1}{3} < \frac{2}{3}$
3. $\frac{5}{8} < \frac{6}{8}$

4. $\frac{3}{10} < \frac{8}{10}$
5. $\frac{1}{4} < \frac{3}{4}$
6. $\frac{6}{7} > \frac{3}{7}$

7. $\frac{4}{6} > \frac{1}{6}$
8. $\frac{5}{9} > \frac{4}{9}$
9. $\frac{12}{24} = \frac{2}{4}$

10. $\frac{1}{2} > \frac{24}{50}$
11. $\frac{4}{5} < \frac{5}{6}$
12. $\frac{9}{12} > \frac{15}{24}$

13. $\frac{1}{2} < \frac{3}{4}$
14. $\frac{1}{6} < \frac{2}{3}$
15. $\frac{3}{4} > \frac{1}{8}$

16. $\frac{2}{4} = \frac{1}{2}$
17. $\frac{6}{8} > \frac{2}{8}$
18. $\frac{1}{3} > \frac{2}{9}$

19. $\frac{4}{6} = \frac{2}{3}$
20. $\frac{1}{5} > \frac{2}{15}$
21. $\frac{3}{4} > \frac{5}{8}$

Answer Key

© Carson-Dellosa • CD-104629

Panel 1 (page 53)

Name _____

4.NF.A.1, 4.NF.B.3a

Adding Fractions with Like Denominators

To add fractions with the same denominators, add the numerators.
Reduce the answer to simplest form.

$\frac{3}{10} + \frac{3}{10} = \frac{6}{10}$

$\frac{6}{10} \div \frac{2}{2} = \frac{3}{5}$

Solve each problem. Write the answer in simplest form.

1. $\frac{2}{5} + \frac{3}{5} = 1$

2. $\frac{2}{7} + \frac{4}{7} = \frac{6}{7}$

3. $\frac{3}{8} + \frac{1}{8} = \frac{1}{2}$

4. $\frac{2}{9} + \frac{3}{9} = \frac{5}{9}$

5. $\frac{5}{8} + \frac{2}{8} = \frac{7}{8}$

6. $\frac{3}{18} + \frac{1}{18} = \frac{2}{9}$

7. $\frac{2}{5} + \frac{2}{5} = \frac{4}{5}$

8. $\frac{1}{8} + \frac{1}{8} = \frac{1}{4}$

9. $\frac{3}{4} + \frac{1}{4} = 1$

10. $\frac{11}{15} + \frac{10}{15} = 1\frac{6}{15}$

11. $\frac{2}{25} + \frac{2}{25} = \frac{4}{25}$

12. $\frac{3}{12} + \frac{1}{12} = \frac{1}{3}$

Panel 2 (page 54)

Name _____

4.NF.A.1, 4.NF.B.3a

Adding Fractions with Like Denominators

Solve each problem. Write the answer in simplest form.

1. $\frac{1}{3} + \frac{2}{3} = 1$

2. $\frac{2}{9} + \frac{5}{9} = \frac{7}{9}$

3. $\frac{1}{6} + \frac{1}{6} = \frac{1}{3}$

4. $\frac{3}{6} + \frac{1}{6} = \frac{2}{3}$

5. $\frac{2}{4} + \frac{2}{4} = 1$

6. $\frac{1}{2} + \frac{1}{2} = 1$

7. $\frac{5}{8} + \frac{3}{8} = 1$

8. $\frac{1}{5} + \frac{2}{5} = \frac{3}{5}$

9. $\frac{2}{10} + \frac{4}{10} = \frac{3}{5}$

10. $\frac{1}{4} + \frac{1}{4} = \frac{1}{2}$

11. $\frac{3}{5} + \frac{2}{5} = 1$

12. $\frac{3}{7} + \frac{2}{7} = \frac{5}{7}$

13. $\frac{2}{3} + \frac{2}{3} = 1\frac{1}{3}$

14. $\frac{1}{7} + \frac{1}{7} = \frac{2}{7}$

15. $\frac{1}{6} + \frac{4}{6} = \frac{5}{6}$

Panel 3 (page 55)

Name _____

4.NF.A.1, 4.NF.B.3a

Adding Fractions with Like Denominators

Solve each problem. Write the answer in simplest form.

1. $\frac{2}{7} + \frac{3}{7} = \frac{5}{7}$

2. $\frac{6}{8} + \frac{1}{8} = \frac{7}{8}$

3. $\frac{7}{10} + \frac{3}{10} = 1$

4. $\frac{3}{7} + \frac{1}{7} = \frac{4}{7}$

5. $\frac{1}{5} + \frac{3}{5} = \frac{4}{5}$

6. $\frac{3}{5} + \frac{2}{5} = 1$

7. $\frac{1}{4} + \frac{2}{4} = \frac{3}{4}$

8. $\frac{1}{5} + \frac{3}{5} = \frac{4}{5}$

9. $\frac{4}{8} + \frac{2}{8} = \frac{3}{4}$

10. $\frac{4}{7} + \frac{5}{7} = 1\frac{2}{7}$

11. $\frac{1}{8} + \frac{5}{8} = \frac{3}{4}$

12. $\frac{3}{8} + \frac{4}{8} = \frac{7}{8}$

13. $\frac{2}{10} + \frac{4}{10} = \frac{3}{5}$

14. $\frac{3}{4} + \frac{1}{4} = 1$

15. $\frac{1}{3} + \frac{1}{3} = \frac{2}{3}$

16. $\frac{8}{9} + \frac{8}{9} = 1\frac{7}{9}$

17. $\frac{2}{6} + \frac{1}{6} = \frac{1}{2}$

18. $\frac{5}{12} + \frac{5}{12} = \frac{5}{6}$

19. $\frac{1}{6} + \frac{3}{6} = \frac{2}{3}$

20. $\frac{2}{9} + \frac{1}{9} = \frac{1}{3}$

Panel 4 (page 56)

Name _____

4.NF.A.1, 4.NF.B.3a

Subtracting Fractions with Like Denominators

To subtract fractions with the same denominators, subtract the numerators.
Reduce the answer to simplest form.

$\frac{8}{10} - \frac{3}{10} = \frac{5}{10}$

$\frac{5}{10} \div \frac{5}{5} = \frac{1}{2}$

Solve each problem. Write the answer in simplest form.

1. $\frac{2}{5} - \frac{1}{5} = \frac{1}{5}$

2. $\frac{6}{7} - \frac{4}{7} = \frac{2}{7}$

3. $\frac{3}{8} - \frac{1}{8} = \frac{1}{4}$

4. $\frac{5}{9} - \frac{3}{9} = \frac{2}{9}$

5. $\frac{5}{8} - \frac{2}{8} = \frac{3}{8}$

6. $\frac{7}{8} - \frac{1}{8} = \frac{3}{4}$

7. $\frac{4}{5} - \frac{2}{5} = \frac{2}{5}$

8. $\frac{4}{8} - \frac{1}{8} = \frac{3}{8}$

9. $\frac{3}{4} - \frac{1}{4} = \frac{1}{2}$

10. $\frac{11}{15} - \frac{10}{15} = \frac{1}{15}$

11. $\frac{3}{5} - \frac{2}{5} = \frac{1}{5}$

12. $\frac{2}{4} - \frac{1}{4} = \frac{1}{4}$

© Carson-Dellosa • CD-104629

Answer Key

Name _____ 4.NF.A.1, 4.NF.B.3a

Subtracting Fractions with Like Denominators
Solve each problem. Write the answer in simplest form.

1. $\frac{2}{5} - \frac{1}{5} = \frac{1}{5}$

2. $\frac{5}{9} - \frac{2}{9} = \frac{1}{3}$

3. $\frac{6}{7} - \frac{1}{7} = \frac{5}{7}$

4. $\frac{5}{6} - \frac{3}{6} = \frac{1}{3}$

5. $\frac{2}{9} - \frac{2}{9} = 0$

6. $\frac{7}{9} - \frac{3}{9} = \frac{4}{9}$

7. $\frac{5}{10} - \frac{2}{10} = \frac{3}{10}$

8. $\frac{5}{5} - \frac{2}{5} = \frac{3}{5}$

9. $\frac{9}{20} - \frac{2}{20} = \frac{7}{20}$

10. $\frac{2}{2} - \frac{1}{2} = \frac{1}{2}$

11. $\frac{3}{8} - \frac{2}{8} = \frac{1}{8}$

12. $\frac{2}{3} - \frac{1}{3} = \frac{1}{3}$

13. $\frac{3}{4} - \frac{2}{4} = \frac{1}{4}$

14. $\frac{1}{7} - \frac{1}{7} = 0$

15. $\frac{9}{10} - \frac{7}{10} = \frac{1}{5}$

57

Name _____ 4.NF.A.1, 4.NF.B.3a

Subtracting Fractions with Like Denominators
Solve each problem. Write the answer in simplest form.

1. $\frac{5}{6} - \frac{1}{6} = \frac{2}{3}$
2. $\frac{7}{12} - \frac{5}{12} = \frac{1}{6}$
3. $\frac{9}{14} - \frac{1}{14} = \frac{4}{7}$
4. $\frac{7}{8} - \frac{5}{8} = \frac{1}{4}$
5. $\frac{6}{8} - \frac{3}{8} = \frac{3}{8}$

6. $\frac{5}{7} - \frac{2}{7} = \frac{3}{7}$
7. $\frac{9}{11} - \frac{1}{11} = \frac{8}{11}$
8. $\frac{5}{9} - \frac{4}{9} = \frac{1}{9}$
9. $\frac{3}{10} - \frac{1}{10} = \frac{1}{5}$
10. $\frac{7}{9} - \frac{1}{9} = \frac{2}{3}$

11. $\frac{5}{8} - \frac{1}{8} = \frac{1}{2}$
12. $\frac{5}{7} - \frac{3}{7} = \frac{2}{7}$
13. $\frac{15}{16} - \frac{11}{16} = \frac{1}{4}$
14. $\frac{4}{5} - \frac{2}{5} = \frac{2}{5}$
15. $\frac{2}{3} - \frac{1}{3} = \frac{1}{3}$

16. $\frac{2}{5} - \frac{1}{5} = \frac{1}{5}$
17. $\frac{3}{4} - \frac{1}{4} = \frac{1}{2}$
18. $\frac{13}{15} - \frac{11}{15} = \frac{2}{15}$
19. $\frac{9}{10} - \frac{7}{10} = \frac{1}{5}$
20. $\frac{3}{3} - \frac{1}{3} = \frac{2}{3}$

58

Name _____ 4.NF.A.1, 4.NF.B.3c

Adding Mixed Numbers with Like Denominators

To add mixed numbers, add the whole numbers and fractions separately:
$$3\frac{1}{3} + 2\frac{1}{3} = (3 + 2) + \left(\frac{1}{3} + \frac{1}{3}\right) = 5\frac{2}{3}$$
OR
Write the mixed numbers as improper fractions and add. Then, change the answer back into a mixed number:
$$3\frac{1}{3} + 2\frac{1}{3} = \frac{10}{3} + \frac{7}{3} = \frac{17}{3} = 5\frac{2}{3}$$

Solve. Add the whole numbers and fractions separately.

1. $1\frac{1}{4} + 2\frac{2}{4} = (\underline{1} + \underline{2}) + (\frac{1}{4} + \frac{2}{4}) = 3\frac{3}{4}$

2. $3\frac{5}{8} + 4\frac{2}{8} = (\underline{3} + \underline{4}) + (\frac{5}{8} + \frac{2}{8}) = 7\frac{7}{8}$

3. $5\frac{2}{6} + 7\frac{3}{6} = (\underline{5} + \underline{7}) + (\frac{2}{6} + \frac{3}{6}) = 12\frac{5}{6}$

4. $8\frac{1}{12} + 9\frac{5}{12} = (\underline{8} + \underline{9}) + (\frac{1}{12} + \frac{5}{12}) = 17\frac{6}{12}$

Solve. Write the mixed numbers as improper fractions and add. Show your answer as a mixed number in simplest form.

5. $4\frac{3}{7} + 3\frac{2}{7} = \frac{31}{7} + \frac{23}{7} = \frac{54}{7} = 7\frac{5}{7}$

6. $6\frac{2}{8} + 5\frac{4}{8} = \frac{50}{8} + \frac{44}{8} = \frac{94}{8} = 11\frac{3}{4}$

7. $3\frac{4}{10} + 1\frac{1}{10} = \frac{34}{10} + \frac{11}{10} = \frac{45}{10} = 4\frac{1}{2}$

8. $4\frac{2}{5} + 2\frac{1}{5} = \frac{22}{5} + \frac{11}{5} = \frac{33}{5} = 6\frac{3}{5}$

59

Name _____ 4.NF.A.1, 4.NF.B.3c

Adding Mixed Numbers with Like Denominators

Remember: If the fractional parts add up to more than one whole, add the whole to the whole number part of the answer.

Solve each problem. Write the answer in simplest form.

1. $2\frac{2}{5} + 2\frac{3}{5} = 5$

2. $5\frac{2}{7} + 6\frac{4}{7} = 11\frac{6}{7}$

3. $3\frac{3}{8} + 4\frac{1}{8} = 7\frac{1}{2}$

4. $4\frac{2}{9} + 5\frac{3}{9} = 9\frac{5}{9}$

5. $6\frac{5}{8} + 7\frac{2}{8} = 13\frac{7}{8}$

6. $5\frac{3}{8} + 4\frac{2}{8} = 9\frac{5}{8}$

7. $8\frac{2}{5} + 1\frac{2}{5} = 9\frac{4}{5}$

8. $7\frac{1}{8} + 7\frac{1}{8} = 14\frac{1}{4}$

9. $2\frac{3}{4} + 2\frac{1}{4} = 5$

10. $5\frac{11}{15} + 6\frac{4}{15} = 12$

11. $2\frac{2}{5} + 2\frac{2}{5} = 4\frac{4}{5}$

12. $4\frac{3}{4} + 1\frac{1}{4} = 6$

60

Name _____   4.NF.A.1, 4.NF.B.3c

Adding Mixed Numbers with Like Denominators
Solve each problem. Write the answer in simplest form.

1. $4\frac{5}{8} + 5\frac{4}{8} = 10\frac{1}{8}$
2. $2\frac{2}{5} + 6\frac{4}{5} = 9\frac{1}{5}$
3. $4\frac{5}{8} + 5\frac{4}{8} = 10\frac{1}{8}$
4. $10\frac{3}{4} + 8\frac{2}{4} = 19\frac{1}{4}$

5. $1\frac{1}{2} + 4\frac{1}{2} = 6$
6. $8\frac{4}{9} + 1\frac{5}{9} = 10$
7. $7\frac{6}{9} + 2\frac{1}{9} = 9\frac{7}{9}$
8. $2\frac{5}{6} + 8\frac{5}{6} = 11\frac{2}{3}$

9. $6\frac{2}{3} + 7\frac{2}{3} = 14\frac{1}{3}$
10. $4\frac{2}{7} + 5\frac{3}{7} = 9\frac{5}{7}$
11. $4\frac{2}{5} + 6\frac{4}{5} = 11\frac{1}{5}$
12. $9\frac{4}{12} + 6\frac{10}{12} = 16\frac{1}{6}$

13. $3\frac{9}{10} + 7\frac{6}{10} = 11\frac{1}{2}$
14. $3\frac{1}{3} + 4\frac{2}{3} = 8$
15. $3\frac{1}{3} + 5\frac{2}{3} = 9$
16. $1\frac{4}{5} + 5\frac{3}{5} = 7\frac{2}{5}$

Name _____   4.NF.A.1, 4.NF.B.3c

Subtracting Mixed Numbers with Like Denominators

> To subtract mixed numbers, subtract the whole numbers and fractions separately:
> $$5\frac{5}{7} - 3\frac{1}{7} = (5-3) + (\frac{5}{7} - \frac{1}{7}) = 2\frac{4}{7}$$
> OR
> Write the mixed numbers as improper fractions and subtract. Then, change the answer back into a mixed number:
> $$5\frac{5}{7} - 3\frac{1}{7} = \frac{40}{7} - \frac{22}{7} = \frac{18}{7} = 2\frac{4}{7}$$

Solve. Subtract the whole numbers and fractions separately.

1. $5\frac{5}{8} - 3\frac{4}{8} = (5 - 3) + (\frac{5}{8} - \frac{4}{8}) = 2\frac{1}{8}$
2. $7\frac{3}{5} - 5\frac{1}{5} = (7 - 5) + (\frac{3}{5} - \frac{1}{5}) = 2\frac{2}{5}$
3. $2\frac{1}{6} - 1\frac{1}{6} = (2 - 1) + (\frac{1}{6} - \frac{1}{6}) = 1$
4. $4\frac{9}{10} - 2\frac{7}{10} = (4 - 2) + (\frac{9}{10} - \frac{7}{10}) = 2\frac{1}{5}$

Solve. Write the mixed numbers as improper fractions and subtract. Show answer as a mixed number in simplest form.

5. $7\frac{2}{3} - 3\frac{1}{3} = \frac{23}{3} - \frac{10}{3} = \frac{13}{3} = 4\frac{1}{3}$
6. $6\frac{7}{8} - 1\frac{1}{8} = \frac{55}{8} - \frac{9}{8} = \frac{46}{8} = 5\frac{3}{4}$
7. $8\frac{4}{10} - 4\frac{2}{10} = \frac{84}{10} - \frac{42}{10} = \frac{42}{10} = 4\frac{1}{5}$
8. $9\frac{6}{7} - 2\frac{4}{7} = \frac{69}{7} - \frac{18}{7} = \frac{51}{7} = 7\frac{2}{7}$

Name _____   4.NF.A.1, 4.NF.B.3c

Subtracting Mixed Numbers with Like Denominators

> If the numerator of the first number is smaller than the numerator of the second number, borrow from the whole number. Add the whole to the fraction to create an improper fraction. Then, subtract.
>
> $6\frac{2}{5}$ → $5\frac{7}{5}$ Borrow 1 from the 6 as $\frac{5}{5}$ and add to $\frac{2}{5}$.
> $-3\frac{4}{5}$ $-3\frac{4}{5}$
> _____   _____
> $2\frac{3}{5}$

Solve each problem. Write the answer in simplest form.

1. $5\frac{5}{8} - 2\frac{4}{8} = 3\frac{1}{8}$
2. $6\frac{2}{6} - 3\frac{5}{6} = 2\frac{1}{2}$
3. $5\frac{5}{8} - 3\frac{4}{8} = 2\frac{1}{8}$
4. $7\frac{3}{5} - 5\frac{4}{5} = 1\frac{4}{5}$

5. $8\frac{4}{5} - 4\frac{1}{5} = 4\frac{3}{5}$
6. $3\frac{2}{6} - 2\frac{1}{6} = 1\frac{1}{6}$
7. $9\frac{3}{7} - 2\frac{5}{7} = 6\frac{5}{7}$
8. $8\frac{5}{9} - 3\frac{6}{9} = 4\frac{8}{9}$

9. $5\frac{1}{3} - 1\frac{2}{3} = 3\frac{2}{3}$
10. $10\frac{1}{4} - 7\frac{3}{4} = 2\frac{1}{2}$
11. $5\frac{3}{6} - 4\frac{2}{6} = 1\frac{1}{3}$
12. $4\frac{9}{10} - 2\frac{7}{10} = 2\frac{1}{5}$

Name _____   4.NF.A.1, 4.NF.B.3c

Subtracting Mixed Numbers with Like Denominators
Solve each problem. Write the answer in simplest form.

1. $12\frac{7}{8} - 5\frac{5}{8} = 7\frac{1}{4}$
2. $2\frac{2}{3} - 2\frac{1}{3} = \frac{1}{3}$
3. $9\frac{7}{8} - 4\frac{4}{8} = 5\frac{3}{8}$
4. $3\frac{1}{8} - 1\frac{7}{8} = 1\frac{1}{4}$

5. $10\frac{1}{5} - 7\frac{4}{5} = 2\frac{2}{5}$
6. $3\frac{1}{4} - 2\frac{3}{4} = \frac{1}{2}$
7. $10\frac{2}{3} - 9\frac{1}{3} = 1\frac{1}{3}$
8. $5\frac{4}{5} - 4\frac{1}{5} = 1\frac{3}{5}$

9. $5\frac{1}{3} - 1\frac{2}{3} = 3\frac{2}{3}$
10. $8\frac{7}{10} - 7\frac{9}{10} = \frac{4}{5}$
11. $8\frac{3}{16} - 7\frac{5}{16} = \frac{7}{8}$
12. $6\frac{7}{15} - 2\frac{8}{15} = 3\frac{14}{15}$

13. $6\frac{2}{12} - 3\frac{2}{12} = 3$
14. $4\frac{5}{6} - 2\frac{1}{6} = 2\frac{2}{3}$
15. $4\frac{11}{18} - 1\frac{7}{18} = 3\frac{2}{9}$
16. $8\frac{3}{10} - 1\frac{7}{10} = 6\frac{3}{5}$

Answer Key

Name _____   4.NF.B.4

Multiplying Fractions by Whole Numbers

Fractions can be used to identify part of a set.

  There are 6 circles. One-half $(\frac{1}{2})$ of the circles are shaded. $\frac{1}{2}$ of 6 = 3

If you do not have a picture to find a fraction of a set, use division to help.

To find $\frac{1}{2}$ of 6, divide the whole number, 6, by the denominator, 2.
Multiply the quotient, 3, by the numerator, 1.

$6 \div 2 = 3$ $3 \times 1 = 3$ $\frac{1}{2}$ of 6 = 3

To find $\frac{2}{3}$ of 24, divide the whole number, 24, by the denominator, 3.
Multiply the quotient, 8, by the numerator, 2.

$24 \div 3 = 8$ $8 \times 2 = 16$ $\frac{2}{3}$ of 24 = 16

Solve each problem.

1. $\frac{2}{11}$ of 44 = **8**

2. $\frac{2}{7}$ of 49 = **14**

3. $\frac{2}{4}$ of 36 = **18**

4. $\frac{4}{6}$ of 48 = **32**

5. $\frac{3}{9}$ of 81 = **27**

6. $\frac{3}{12}$ of 24 = **6**

7. $\frac{2}{3}$ of 33 = **22**

8. $\frac{3}{4}$ of 20 = **15**

9. $\frac{1}{3}$ of 12 = **4**

65

Name _____   4.NF.B.4

Multiplying Fractions by Whole Numbers

To multiply fractions, change the whole number to a fraction with a denominator of 1. Then, multiply the numerators and multiply the denominators. Last, simplify.

Solve each problem. Write the answer in simplest form.

1. $4 \times \frac{1}{2} =$ **2**

2. $2 \times \frac{2}{5} = \frac{\mathbf{4}}{\mathbf{5}}$

3. $4 \times \frac{2}{7} = \mathbf{1}\frac{\mathbf{1}}{\mathbf{7}}$

4. $3 \times \frac{5}{6} = \mathbf{2}\frac{\mathbf{1}}{\mathbf{2}}$

5. $8 \times \frac{1}{8} =$ **1**

6. $\frac{2}{15} \times 3 = \frac{\mathbf{2}}{\mathbf{5}}$

7. $\frac{1}{8} \times 5 = \frac{\mathbf{5}}{\mathbf{8}}$

8. $\frac{5}{7} \times 5 = \mathbf{3}\frac{\mathbf{4}}{\mathbf{7}}$

9. $\frac{2}{3} \times 2 = \mathbf{1}\frac{\mathbf{1}}{\mathbf{3}}$

10. $\frac{3}{16} \times 4 = \frac{\mathbf{3}}{\mathbf{4}}$

11. $\frac{1}{3} \times 7 = \mathbf{2}\frac{\mathbf{1}}{\mathbf{3}}$

12. $4 \times \frac{3}{4} =$ **3**

13. $\frac{6}{8} \times 2 = \mathbf{1}\frac{\mathbf{1}}{\mathbf{2}}$

14. $5 \times \frac{4}{5} =$ **4**

15. $3 \times \frac{2}{3} =$ **2**

66

Name _____   4.NF.B.4

Multiplying Fractions by Whole Numbers

Solve each problem. Write the answer in simplest form.

1. $9 \times \frac{2}{3} =$ **6**

2. $14 \times \frac{4}{7} =$ **8**

3. $77 \times \frac{10}{11} =$ **70**

4. $36 \times \frac{2}{288} = \frac{\mathbf{1}}{\mathbf{4}}$

5. $16 \times \frac{4}{8} =$ **8**

6. $9 \times \frac{5}{6} = \mathbf{7}\frac{\mathbf{1}}{\mathbf{2}}$

7. $3 \times \frac{1}{3} =$ **1**

8. $30 \times \frac{3}{90} =$ **1**

9. $12 \times \frac{1}{36} = \frac{\mathbf{1}}{\mathbf{3}}$

10. $5 \times \frac{2}{5} =$ **2**

11. $12 \times \frac{7}{8} = \mathbf{10}\frac{\mathbf{1}}{\mathbf{2}}$

12. $5 \times \frac{3}{40} = \frac{\mathbf{3}}{\mathbf{8}}$

13. $22 \times \frac{1}{44} = \frac{\mathbf{1}}{\mathbf{2}}$

14. $4 \times \frac{1}{8} = \frac{\mathbf{1}}{\mathbf{2}}$

15. $81 \times \frac{2}{3} =$ **54**

67

Name _____   4.NF.C.6

Relating Decimals and Fractions

A **decimal** is a number that uses a **decimal point** (.) to show tenths and hundredths instead of a fraction.

A **tenth** is one out of 10 equal parts of a whole.

A **hundredth** is one out of 100 equal parts of a whole.

 1 or 1.0 $\frac{6}{10}$ or 0.6 $1\frac{6}{10}$ 1.6

 $\frac{100}{100}$ or 1.0 $\frac{47}{100}$ or 0.47 $1\frac{47}{100}$ 1.47

Write two ways to name each picture.

1. Fraction: $\frac{\mathbf{2}}{\mathbf{10}}$ Decimal: **0.2**

2. Fraction: $\mathbf{1}\frac{\mathbf{4}}{\mathbf{10}}$ Decimal: **1.4**

3. Fraction: $\mathbf{1}\frac{\mathbf{1}}{\mathbf{10}}$ Decimal: **1.1**

4. Fraction: $\frac{\mathbf{34}}{\mathbf{100}}$ Decimal: **0.34**

5. Fraction: $\frac{\mathbf{69}}{\mathbf{100}}$ Decimal: **0.69**

6. Fraction: $\mathbf{1}\frac{\mathbf{7}}{\mathbf{100}}$ Decimal: **1.07**

68

Answer Key

Relating Decimals and Fractions
Write each decimal as a fraction in simplest form.

1. $0.5 = \dfrac{1}{2}$ 2. $0.9 = \dfrac{9}{10}$ 3. $0.7 = \dfrac{7}{10}$

4. $9.5 = 9\dfrac{1}{2}$ 5. $1.8 = 1\dfrac{4}{5}$ 6. $2.2 = 2\dfrac{1}{5}$

7. $0.62 = \dfrac{31}{50}$ 8. $1.25 = 1\dfrac{1}{4}$ 9. $0.1 = \dfrac{1}{10}$

10. $0.22 = \dfrac{11}{50}$ 11. $4.10 = 4\dfrac{1}{10}$ 12. $0.36 = \dfrac{9}{25}$

Write each fraction as a decimal.

13. $\dfrac{3}{10} = 0.3$ 14. $\dfrac{9}{10} = 0.9$ 15. $8\dfrac{8}{10} = 8.8$

16. $2\dfrac{5}{10} = 2.5$ 17. $\dfrac{4}{10} = 0.4$ 18. $\dfrac{88}{100} = 0.88$

19. $\dfrac{52}{100} = 0.52$ 20. $3\dfrac{25}{100} = 3.25$ 21. $6\dfrac{5}{10} = 6.5$

22. $\dfrac{14}{100} = 0.14$ 23. $4\dfrac{1}{10} = 4.1$ 24. $9\dfrac{30}{100} = 9.30$

Relating Decimals and Fractions
Write as a decimal or a fraction in simplest form.

1. $8.2 = 8\dfrac{1}{5}$ 2. $5.4 = 5\dfrac{2}{5}$ 3. $48\dfrac{2}{10} = 48.2$

4. $0.15 = \dfrac{3}{20}$ 5. $25.32 = 25\dfrac{8}{25}$ 6. $3.25 = 3\dfrac{1}{4}$

7. $30.2 = 30\dfrac{1}{5}$ 8. $\dfrac{65}{100} = 0.65$ 9. $9.1 = 9\dfrac{1}{10}$

10. $10.6 = 10\dfrac{3}{5}$ 11. $\dfrac{20}{100} = 0.20$ 12. $\dfrac{2}{10} = 0.2$

13. $86.12 = 86\dfrac{3}{25}$ 14. $6.5 = 6\dfrac{1}{2}$ 15. $9\dfrac{9}{10} = 9.9$

16. $1\dfrac{29}{100} = 1.29$ 17. $7.6 = 7\dfrac{2}{3}$ 18. $\dfrac{99}{100} = 0.99$

19. $0.75 = \dfrac{3}{4}$ 20. $4.36 = 4\dfrac{9}{25}$ 21. $9.45 = 9\dfrac{9}{20}$

22. $75\dfrac{2}{100} = 75.02$ 23. $25.2 = 25\dfrac{1}{5}$ 24. $20\dfrac{6}{10} = 20.6$

Comparing Decimals

> To compare decimals, first look at the whole numbers.
>
> $2.68 < 4.43$ $48.52 > 12.71$
>
> If the whole numbers are the same, compare the tenths.
>
> $7.83 > 7.38$ $80.74 > 80.07$
>
> If the whole number and the tenths are the same, compare the hundredths.
>
> $54.19 > 54.12$ $3.40 < 3.44$

Use > or < to compare the decimals.

1. $9.4 < 9.9$ 2. $4.7 > 4.2$ 3. $34.93 < 49.04$

4. $8.68 > 8.30$ 5. $2.8 > 2.3$ 6. $17.45 < 17.46$

7. $22.31 > 22.18$ 8. $1.4 < 1.7$ 9. $5.6 < 5.8$

10. $35.73 < 35.81$ 11. $11.1 > 1.11$ 12. $3.97 < 3.99$

13. $9.14 < 91.4$ 14. $35.1 < 55.9$ 15. $6.4 > 4.9$

Comparing Decimals

> When comparing decimals, first compare the whole numbers. Then, compare the digits in the tenths columns. If the digits in the tenths columns are the same, compare the digits in the hundredths columns.

Use > or < to compare the decimals.

1. $0.6 > 0.4$ 2. $0.1 < 0.5$ 3. $0.23 > 0.03$ 4. $0.6 < 0.9$

5. $0.06 < 0.60$ 6. $0.4 < 0.7$ 7. $0.9 > 0.5$ 8. $0.7 > 0.6$

9. $0.42 > 0.14$ 10. $0.72 > 0.27$ 11. $0.25 < 0.52$ 12. $0.7 > 0.3$

13. $1.4 < 1.6$ 14. $3.5 < 3.7$ 15. $16.2 < 16.8$ 16. $5.21 < 5.38$

17. $2.48 > 2.35$ 18. $14.5 > 14.3$ 19. $42.6 > 42.3$ 20. $3.8 < 3.9$

Answer Key

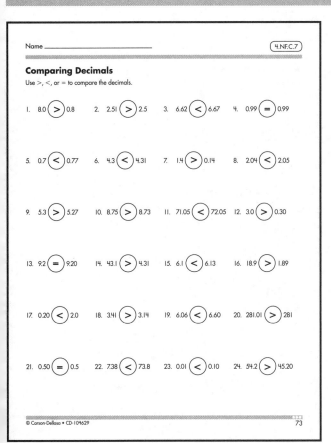

Name _____ (4.NF.C.7)

Comparing Decimals

Use >, <, or = to compare the decimals.

1. 8.0 (>) 0.8   2. 2.51 (>) 2.5   3. 6.62 (<) 6.67   4. 0.99 (=) 0.99

5. 0.7 (<) 0.77   6. 4.3 (<) 4.31   7. 1.4 (>) 0.14   8. 2.04 (<) 2.05

9. 5.3 (>) 5.27   10. 8.75 (>) 8.73   11. 71.05 (<) 72.05   12. 3.0 (>) 0.30

13. 9.2 (=) 9.20   14. 43.1 (>) 4.31   15. 6.1 (<) 6.13   16. 18.9 (>) 1.89

17. 0.20 (<) 2.0   18. 3.41 (>) 3.14   19. 6.06 (<) 6.60   20. 281.01 (>) 281

21. 0.50 (=) 0.5   22. 7.38 (<) 73.8   23. 0.01 (<) 0.10   24. 54.2 (>) 45.20

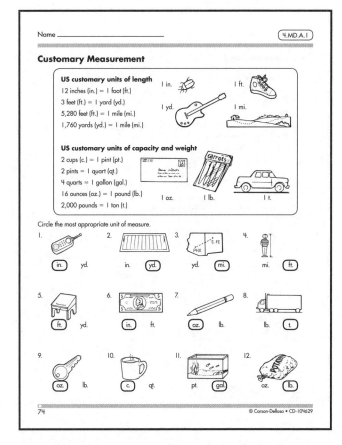

Name _____ (4.MD.A.1)

Customary Measurement

US customary units of length
12 inches (in.) = 1 foot (ft.)
3 feet (ft.) = 1 yard (yd.)
5,280 feet (ft.) = 1 mile (mi.)
1,760 yards (yd.) = 1 mile (mi.)

US customary units of capacity and weight
2 cups (c.) = 1 pint (pt.)
2 pints = 1 quart (qt.)
4 quarts = 1 gallon (gal.)
16 ounces (oz.) = 1 pound (lb.)
2,000 pounds = 1 ton (t.)

Circle the most appropriate unit of measure.

1. in. (yd.)   2. in. (yd.)   3. yd. (mi.)   4. mi. (ft.)

5. (ft.) yd.   6. (in.) ft.   7. (oz.) lb.   8. lb. (t.)

9. (oz.) lb.   10. (c.) qt.   11. pt. (gal.)   12. oz. (lb.)

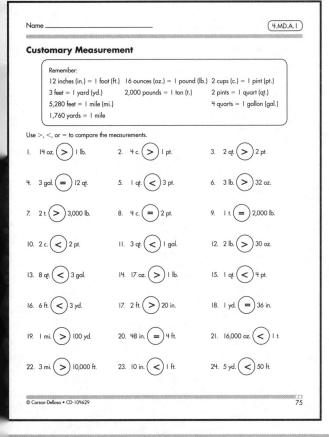

Name _____ (4.MD.A.1)

Customary Measurement

Remember:
12 inches (in.) = 1 foot (ft.)   16 ounces (oz.) = 1 pound (lb.)   2 cups (c.) = 1 pint (pt.)
3 feet = 1 yard (yd.)   2,000 pounds = 1 ton (t.)   2 pints = 1 quart (qt.)
5,280 feet = 1 mile (mi.)   4 quarts = 1 gallon (gal.)
1,760 yards = 1 mile

Use >, <, or = to compare the measurements.

1. 14 oz. (>) 1 lb.   2. 4 c. (>) 1 pt.   3. 2 qt. (>) 2 pt.

4. 3 gal. (=) 12 qt.   5. 1 qt. (<) 3 pt.   6. 3 lb. (>) 32 oz.

7. 2 t. (>) 3,000 lb.   8. 4 c. (=) 2 pt.   9. 1 t. (=) 2,000 lb.

10. 2 c. (<) 2 pt.   11. 3 qt. (<) 1 gal.   12. 2 lb. (>) 30 oz.

13. 8 qt. (<) 3 gal.   14. 17 oz. (>) 1 lb.   15. 1 qt. (<) 4 pt.

16. 6 ft. (<) 3 yd.   17. 2 ft. (>) 20 in.   18. 1 yd. (=) 36 in.

19. 1 mi. (>) 100 yd.   20. 48 in. (=) 4 ft.   21. 16,000 oz. (<) 1 t.

22. 3 mi. (>) 10,000 ft.   23. 10 in. (<) 1 ft.   24. 5 yd. (<) 50 ft.

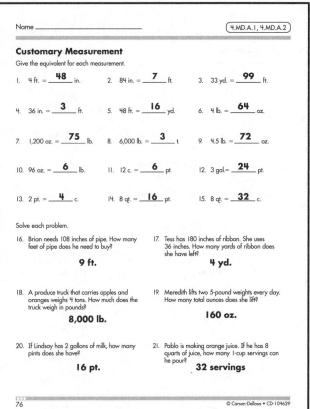

Name _____ (4.MD.A.1, 4.MD.A.2)

Customary Measurement

Give the equivalent for each measurement.

1. 4 ft. = **48** in.   2. 84 in. = **7** ft.   3. 33 yd. = **99** ft.

4. 36 in. = **3** ft.   5. 48 ft. = **16** yd.   6. 4 lb. = **64** oz.

7. 1,200 oz. = **75** lb.   8. 6,000 lb. = **3** t.   9. 4.5 lb. = **72** oz.

10. 96 oz. = **6** lb.   11. 12 c. = **6** pt.   12. 3 gal. = **24** pt.

13. 2 pt. = **4** c.   14. 8 qt. = **16** pt.   15. 8 qt. = **32** c.

Solve each problem.

16. Brian needs 108 inches of pipe. How many feet of pipe does he need to buy?

9 ft.

17. Tess has 180 inches of ribbon. She uses 36 inches. How many yards of ribbon does she have left?

4 yd.

18. A produce truck that carries apples and oranges weighs 4 tons. How much does the truck weigh in pounds?

8,000 lb.

19. Meredith lifts two 5-pound weights every day. How many total ounces does she lift?

160 oz.

20. If Lindsay has 2 gallons of milk, how many pints does she have?

16 pt.

21. Pablo is making orange juice. If he has 8 quarts of juice, how many 1-cup servings can he pour?

32 servings

Answer Key

Panel 1 (page 77)

Name _____   4.MD.A.1

Metric Measurement

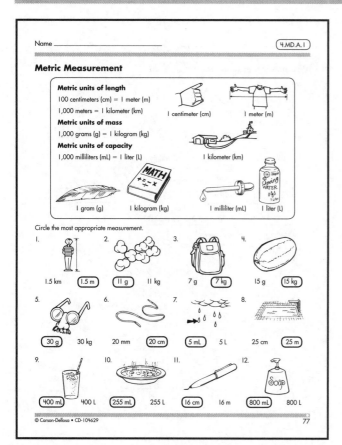

Metric units of length
100 centimeters (cm) = 1 meter (m)
1,000 meters = 1 kilometer (km)
Metric units of mass
1,000 grams (g) = 1 kilogram (kg)
Metric units of capacity
1,000 milliliters (mL) = 1 liter (L)

1 centimeter (cm)   1 meter (m)

1 kilometer (km)

1 gram (g)   1 kilogram (kg)   1 milliliter (mL)   1 liter (L)

Circle the most appropriate measurement.

1. 1.5 km (1.5 m)
2. (11 g) 11 kg
3. 7 g (7 kg)
4. 15 g (15 kg)
5. (30 g) 30 kg
6. 20 mm (20 cm)
7. (5 mL) 5 L
8. 25 cm (25 m)
9. (400 mL) 400 L
10. (255 mL) 255 L
11. (16 cm) 16 m
12. (800 mL) 800 L

© Carson-Dellosa • CD-104629   77

Panel 2 (page 78)

Name _____   4.MD.A.1

Metric Measurement
Use >, <, or = to compare the measurements.

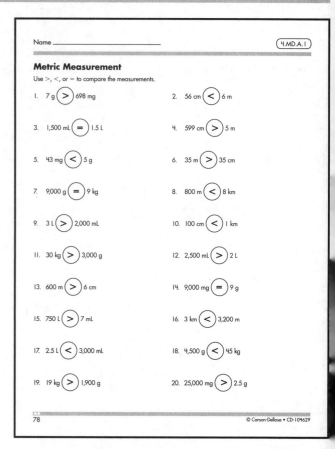

1. 7 g (>) 698 mg
2. 56 cm (<) 6 m
3. 1,500 mL (=) 1.5 L
4. 599 cm (>) 5 m
5. 43 mg (<) 5 g
6. 35 m (>) 35 cm
7. 9,000 g (=) 9 kg
8. 800 m (<) 8 km
9. 3 L (>) 2,000 mL
10. 100 cm (<) 1 km
11. 30 kg (>) 3,000 g
12. 2,500 mL (>) 2 L
13. 600 m (>) 6 cm
14. 9,000 mg (=) 9 g
15. 750 L (>) 7 mL
16. 3 km (<) 3,200 m
17. 2.5 L (<) 3,000 mL
18. 4,500 g (<) 45 kg
19. 19 kg (>) 1,900 g
20. 25,000 mg (>) 2.5 g

78   © Carson-Dellosa • CD-104629

Panel 3 (page 79)

Name _____   4.MD.A.1 , 4.MD.A.2

Metric Measurement
Give the equivalent for each measurement.

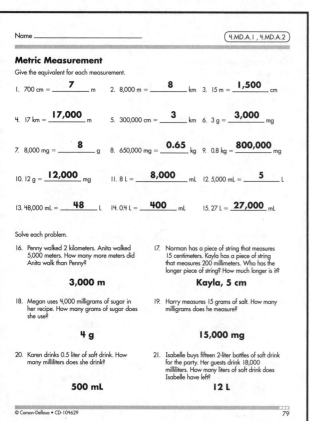

1. 700 cm = **7** m
2. 8,000 m = **8** km
3. 15 m = **1,500** cm
4. 17 km = **17,000** m
5. 300,000 cm = **3** km
6. 3 g = **3,000** mg
7. 8,000 mg = **8** g
8. 650,000 mg = **0.65** kg
9. 0.8 kg = **800,000** mg
10. 12 g = **12,000** mg
11. 8 L = **8,000** mL
12. 5,000 mL = **5** L
13. 48,000 mL = **48** L
14. 0.4 L = **400** mL
15. 27 L = **27,000** mL

Solve each problem.

16. Penny walked 2 kilometers. Anita walked 5,000 meters. How many more meters did Anita walk than Penny?

3,000 m

17. Norman has a piece of string that measures 15 centimeters. Kayla has a piece of string that measures 200 millimeters. Who has the longer piece of string? How much longer is it?

Kayla, 5 cm

18. Megan uses 4,000 milligrams of sugar in her recipe. How many grams of sugar does she use?

4 g

19. Harry measures 15 grams of salt. How many milligrams does he measure?

15,000 mg

20. Karen drinks 0.5 liter of soft drink. How many milliliters does she drink?

500 mL

21. Isabelle buys fifteen 2-liter bottles of soft drink for the party. Her guests drink 18,000 milliliters. How many liters of soft drink does Isabelle have left?

12 L

© Carson-Dellosa • CD-104629   79

Panel 4 (page 80)

Name _____   4.MD.A.3

Finding Perimeter

Perimeter is the total distance around a given figure. To find the perimeter, add the lengths of the sides of the figure.
Example:
    Perimeter = 4 cm + 8 cm + 4 cm + 8 cm
    P = 24 cm

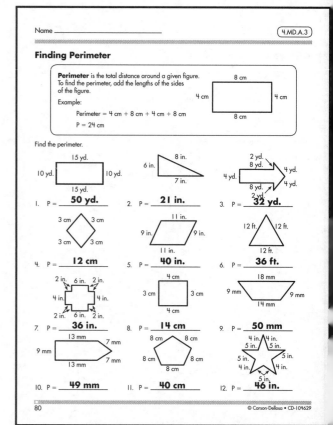

Find the perimeter.

1. P = **50 yd.**
2. P = **21 in.**
3. P = **32 yd.**
4. P = **12 cm**
5. P = **40 in.**
6. P = **36 ft.**
7. P = **36 in.**
8. P = **14 cm**
9. P = **50 mm**
10. P = **49 mm**
11. P = **40 cm**
12. P = **46 in.**

80   © Carson-Dellosa • CD-104629

Answer Key

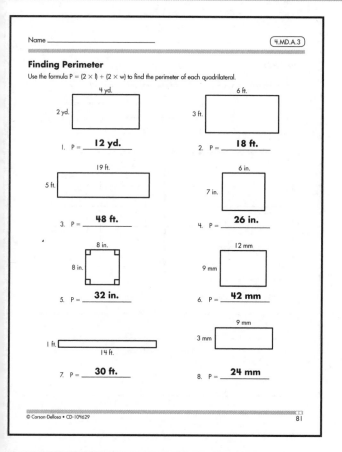

Name _____ (4.MD.A.3)

Finding Perimeter

Use the formula P = (2 × l) + (2 × w) to find the perimeter of each quadrilateral.

1. P = **12 yd.**
2. P = **18 ft.**
3. P = **48 ft.**
4. P = **26 in.**
5. P = **32 in.**
6. P = **42 mm**
7. P = **30 ft.**
8. P = **24 mm**

© Carson-Dellosa • CD-104629   81

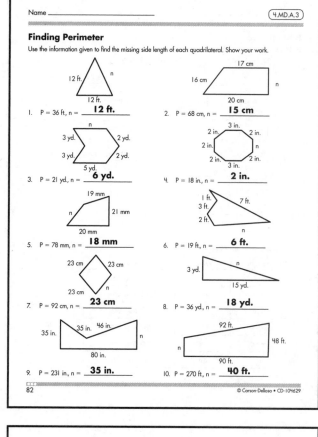

Name _____ (4.MD.A.3)

Finding Perimeter

Use the information given to find the missing side length of each quadrilateral. Show your work.

1. P = 36 ft., n = **12 ft.**
2. P = 68 cm, n = **15 cm**
3. P = 21 yd., n = **6 yd.**
4. P = 18 in., n = **2 in.**
5. P = 78 mm, n = **18 mm**
6. P = 19 ft., n = **6 ft.**
7. P = 92 cm, n = **23 cm**
8. P = 36 yd., n = **18 yd.**
9. P = 231 in., n = **35 in.**
10. P = 270 ft., n = **40 ft.**

82   © Carson-Dellosa • CD-104629

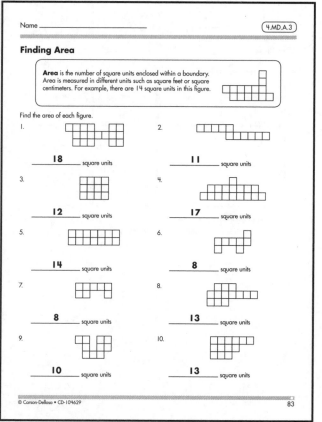

Name _____ (4.MD.A.3)

Finding Area

Area is the number of square units enclosed within a boundary. Area is measured in different units such as square feet or square centimeters. For example, there are 14 square units in this figure.

Find the area of each figure.

1. **18** square units
2. **11** square units
3. **12** square units
4. **17** square units
5. **14** square units
6. **8** square units
7. **8** square units
8. **13** square units
9. **10** square units
10. **13** square units

© Carson-Dellosa • CD-104629   83

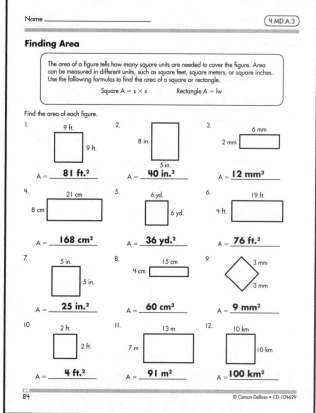

Name _____ (4.MD.A.3)

Finding Area

The area of a figure tells how many square units are needed to cover the figure. Area can be measured in different units, such as square feet, square meters, or square inches. Use the following formulas to find the area of a square or rectangle.

Square A = s × s   Rectangle A = lw

Find the area of each figure.

1. A = **81 ft.²**
2. A = **40 in.²**
3. A = **12 mm²**
4. A = **168 cm²**
5. A = **36 yd.²**
6. A = **76 ft.²**
7. A = **25 in.²**
8. A = **60 cm²**
9. A = **9 mm²**
10. A = **4 ft.²**
11. A = **91 m²**
12. A = **100 km²**

84   © Carson-Dellosa • CD-104629

Answer Key

Finding Area

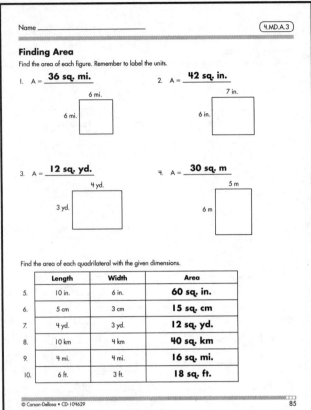

Name _____ (4.MD.A.3)

Finding Area

Find the area of each figure. Remember to label the units.

1. A = **36 sq. mi.**
    6 mi.
    6 mi.

2. A = **42 sq. in.**
    7 in.
    6 in.

3. A = **12 sq. yd.**
    4 yd.
    3 yd.

4. A = **30 sq. m**
    5 m
    6 m

Find the area of each quadrilateral with the given dimensions.

	Length	Width	Area
5.	10 in.	6 in.	**60 sq. in.**
6.	5 cm	3 cm	**15 sq. cm**
7.	4 yd.	3 yd.	**12 sq. yd.**
8.	10 km	4 km	**40 sq. km**
9.	4 mi.	4 mi.	**16 sq. mi.**
10.	6 ft.	3 ft.	**18 sq. ft.**

85

Name _____ (4.MD.B.4)

Line Plots

A **line plot** is a type of graph that shows information on a number line.

Line plots are useful for showing frequency, or the number of times something is repeated.

1. Use a ruler to measure 8 things to the nearest $\frac{1}{2}$ inch. Record your data on the table.

Item	Length	Item	Length

2. Use the data from the table to make a line plot.

- First, look at the data and decide what numbers you will need to include.
- Then, mark each number on the line plot and label it. Do not leave out numbers in between, even if they have no data!
- Finally, mark an X on the line plot to represent each piece of data.

Answers will vary.

86

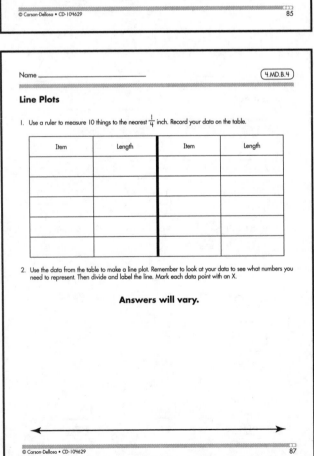

Name _____ (4.MD.B.4)

Line Plots

1. Use a ruler to measure 10 things to the nearest $\frac{1}{4}$ inch. Record your data on the table.

Item	Length	Item	Length

2. Use the data from the table to make a line plot. Remember to look at your data to see what numbers you need to represent. Then divide and label the line. Mark each data point with an X.

Answers will vary.

87

Name _____ (4.MD.B.4)

Line Plots

1. Use a ruler to measure 10 things to the nearest $\frac{1}{8}$ inch. Record your data on the table.

Item	Length	Item	Length

2. Use the data from the table to make a line plot.

Answers will vary.

88

Answer Key

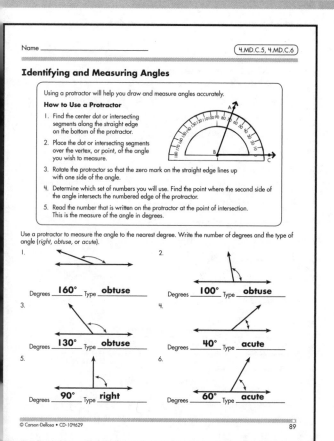

Name _____   4.MD.C.5, 4.MD.C.6

Identifying and Measuring Angles

Using a protractor will help you draw and measure angles accurately.

How to Use a Protractor

1. Find the center dot or intersecting segments along the straight edge on the bottom of the protractor.

2. Place the dot or intersecting segments over the vertex, or point, of the angle you wish to measure.

3. Rotate the protractor so that the zero mark on the straight edge lines up with one side of the angle.

4. Determine which set of numbers you will use. Find the point where the second side of the angle intersects the numbered edge of the protractor.

5. Read the number that is written on the protractor at the point of intersection. This is the measure of the angle in degrees.

Use a protractor to measure the angle to the nearest degree. Write the number of degrees and the type of angle (*right, obtuse,* or *acute*).

1. Degrees **160°** Type **obtuse**
2. Degrees **100°** Type **obtuse**
3. Degrees **130°** Type **obtuse**
4. Degrees **40°** Type **acute**
5. Degrees **90°** Type **right**
6. Degrees **60°** Type **acute**

89

Name _____   4.MD.C.5, 4.MD.C.6

Identifying and Measuring Angles

When measuring angles, remember:

1. Use a ruler to extend the rays if they are too short to measure.

2. Align the base ray on the protractor's straight edge (0°).

3. Read the number the ray intersects. Use the type of angle (acute or obtuse) to help you decide which set of numbers to use.

Measure each angle with a protractor.

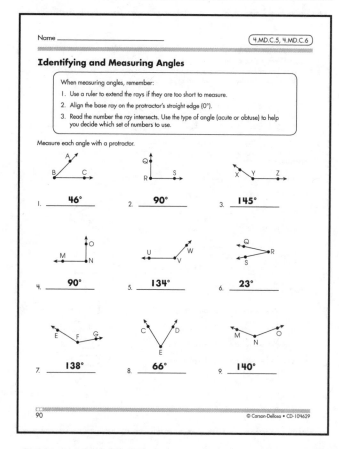

1. **46°**
2. **90°**
3. **145°**
4. **90°**
5. **134°**
6. **23°**
7. **138°**
8. **66°**
9. **140°**

90

Name _____   4.MD.C.5, 4.MD.C.6

Identifying and Measuring Angles

Use a protractor to measure each specified angle to the nearest degree. Write the type (*right, obtuse,* or *acute*) and measure of each angle.

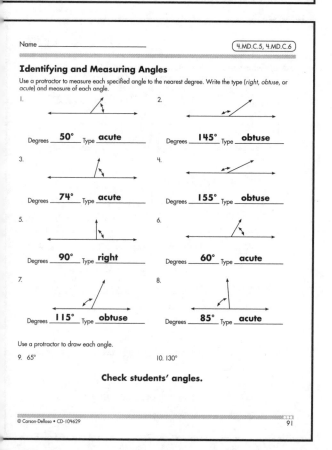

1. Degrees **50°** Type **acute**
2. Degrees **145°** Type **obtuse**
3. Degrees **74°** Type **acute**
4. Degrees **155°** Type **obtuse**
5. Degrees **90°** Type **right**
6. Degrees **60°** Type **acute**
7. Degrees **115°** Type **obtuse**
8. Degrees **85°** Type **acute**

Use a protractor to draw each angle.

9. 65°   10. 130°

Check students' angles.

91

Name _____   4.MD.C.5, 4.G.A.2

Identifying Triangles by Angle

A **triangle** is a three-sided polygon. A triangle's angles can be used to classify it.

Acute Triangle	Equiangular Triangle	Right Triangle	Obtuse Triangle
three acute angles	three congruent angles	one right angle	one obtuse angle

Classify each triangle below by its angles. Write *acute, equiangular, right,* or *obtuse*.

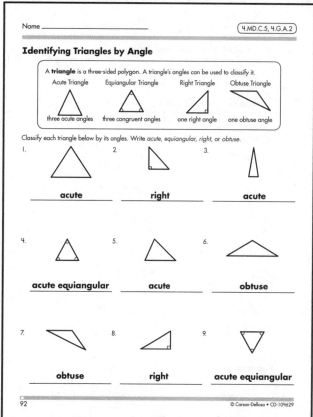

1. **acute**
2. **right**
3. **acute**
4. **acute equiangular**
5. **acute**
6. **obtuse**
7. **obtuse**
8. **right**
9. **acute equiangular**

92

Answer Key

Name _____

4.MD.C.5, 4.G.A.2

Identifying Triangles by Angle

The sum of the three angles of every triangle equals 180°.

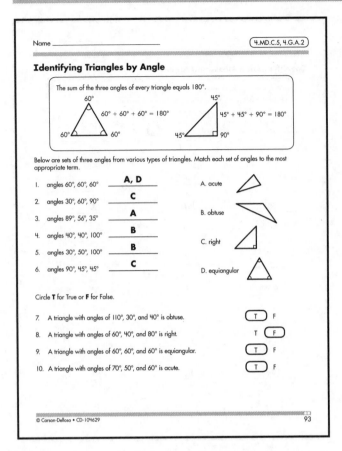

60°
60° + 60° + 60° = 180°
60°   60°

45°
45° + 45° + 90° = 180°
45°   90°

Below are sets of three angles from various types of triangles. Match each set of angles to the most appropriate term.

1. angles 60°, 60°, 60°   **A, D**   A. acute
2. angles 30°, 60°, 90°   **C**
3. angles 89°, 56°, 35°   **A**   B. obtuse
4. angles 40°, 40°, 100°   **B**
5. angles 30°, 50°, 100°   **B**   C. right
6. angles 90°, 45°, 45°   **C**
                                         D. equiangular

Circle **T** for True or **F** for False.

7. A triangle with angles of 110°, 30°, and 40° is obtuse.   (T) F
8. A triangle with angles of 60°, 40°, and 80° is right.   T (F)
9. A triangle with angles of 60°, 60°, and 60° is equiangular. (T) F
10. A triangle with angles of 70°, 50°, and 60° is acute.   (T) F

© Carson-Dellosa • CD-104629   93

Name _____

4.MD.C.5, 4.MD.C.7, 4.G.A.2

Identifying Triangles by Angle

Match each description with the correct triangle names from the box.

1. Angles measure 90°, 100°, 30°. **E**
2. Angles measure 30°, 60°, 90°.   **A**
3. Angles measure 25°, 10°, 145°. **B**
4. Angles measure 15°, 30°, 12°.   **E**
5. Angles measure 45°, 90°, 45°.   **A**
6. Angles measure 60°, 60°, 60°.   **C, D**
7. Angles measure 50°, 88°, 42°.   **C**

A. right
B. obtuse
C. acute
D. equiangular
E. not a possible triangle

Complete each statement using *sometimes, always,* or *never.*

8. An obtuse triangle is _____**never**_____ an equilateral triangle.
9. A right triangle _____**always**_____ has a right angle and two acute angles.
10. An acute triangle is _____**sometimes**_____ an equiangular triangle.

Explain why each shape below is not possible.

11. Right obtuse triangle

        Answers will vary.

12. Triangle with two obtuse angles

        Answers will vary.

94   © Carson-Dellosa • CD-104629

Name _____

4.G.A.1

Identifying Lines, Rays, and Line Segments

A **ray** is a portion of a line that extends from one endpoint infinitely in one direction. The ray to the right is named $\overrightarrow{AB}$, with the endpoint written first and any point on the ray written next.

A **line segment** is a finite portion of a line that contains two endpoints. The segment to the right is named $\overline{AB}$. The segment must be named by its two endpoints.

Identify the following as a *line, ray, line segment,* or *points.*

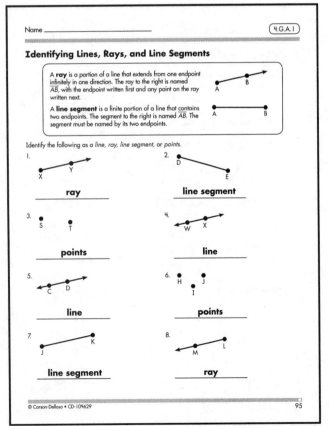

1. **ray**

2. **line segment**

3. **points**

4. **line**

5. **line**

6. **points**

7. **line segment**

8. **ray**

© Carson-Dellosa • CD-104629   95

Name _____

4.G.A.1

Identifying Lines, Rays, and Line Segments

The straight path between points X and Y is a **line segment.** ($\overline{XY}$)

A **line** is a straight path that goes unending in two directions. ($\overleftrightarrow{CD}$)

A **ray** is a straight path that begins at a point and goes unending in one direction. ($\overrightarrow{TX}$)

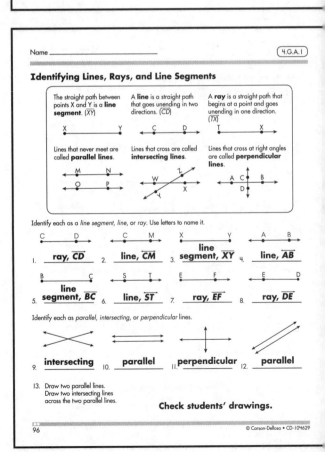

Lines that never meet are called **parallel lines.**

Lines that cross are called **intersecting lines.**

Lines that cross at right angles are called **perpendicular lines.**

Identify each as a *line segment, line,* or *ray.* Use letters to name it.

1. **ray, $\overrightarrow{CD}$**   2. **line, $\overleftrightarrow{CM}$**   3. **line segment, $\overline{XY}$**   4. **line, $\overleftrightarrow{AB}$**

5. **line segment, $\overline{BC}$**   6. **line, $\overleftrightarrow{ST}$**   7. **ray, $\overrightarrow{EF}$**   8. **ray, $\overrightarrow{DE}$**

Identify each as *parallel, intersecting,* or *perpendicular* lines.

9. **intersecting**   10. **parallel**   11. **perpendicular**   12. **parallel**

13. Draw two parallel lines. Draw two intersecting lines across the two parallel lines.

        Check students' drawings.

96   © Carson-Dellosa • CD-104629

Answer Key

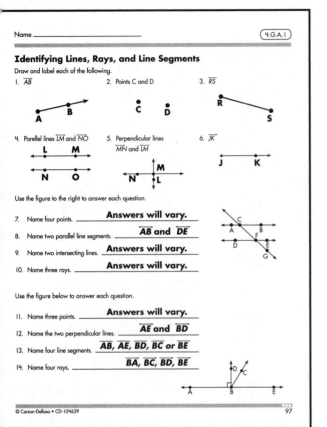

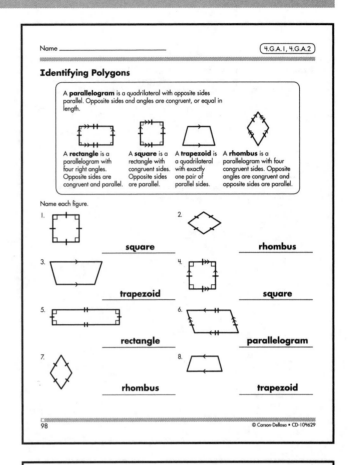

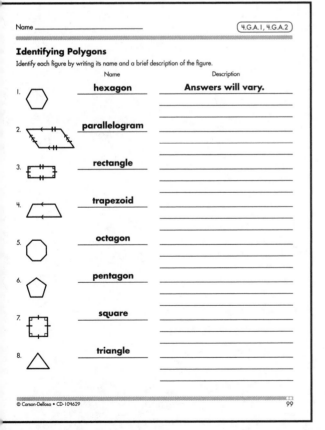

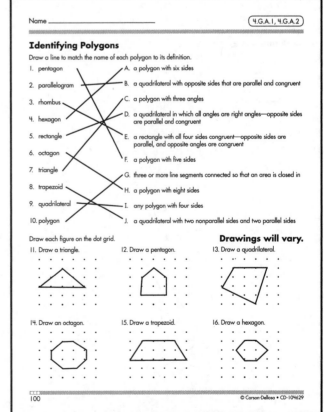

Answer Key

Exploring Line Symmetry

A figure has **line symmetry** if it can be folded along a line so that the two halves are mirror images.

These figures have line symmetry. The heart has one line of symmetry. The rectangle has two lines of symmetry.

These figures do not have line symmetry.

Determine if the following figures have line symmetry and write *yes* or *no*. If yes, draw all of the lines of symmetry.

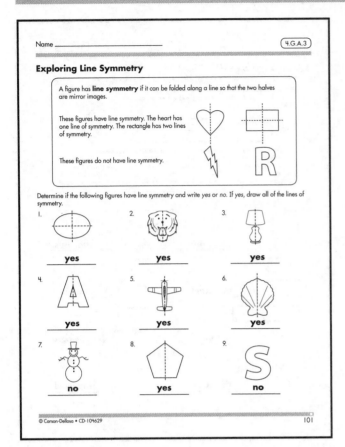

1. yes
2. yes
3. yes
4. yes
5. yes
6. yes
7. no
8. yes
9. no

Exploring Line Symmetry

Some objects have more than one line of symmetry. A regular hexagon has six lines of symmetry.

Decide whether these figures have one line of symmetry, two lines of symmetry, or no lines of symmetry. Write *one*, *two*, or *none*. Then, draw the line or lines.

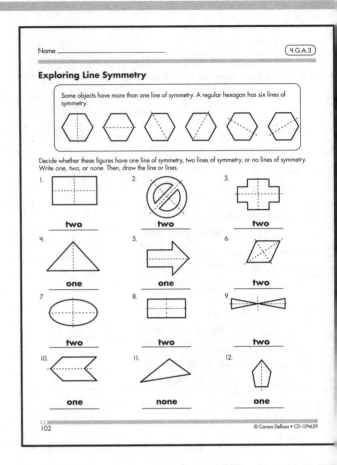

1. two
2. two
3. two
4. one
5. one
6. two
7. two
8. two
9. two
10. one
11. none
12. one

Exploring Line Symmetry

Complete each design to show symmetry. Then, draw the line of symmetry.

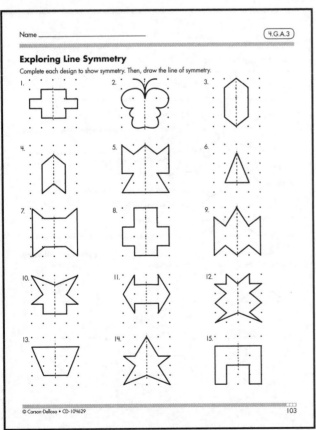

Congratulations!

receives this award for

_____

Signed

Date

$$\begin{array}{r} 38 \\ \times\,12 \\ \hline \end{array}$$

$$\begin{array}{r} 51 \\ \times\,9 \\ \hline \end{array}$$

$$\begin{array}{r} 89 \\ \times\,6 \\ \hline \end{array}$$

$$\begin{array}{r} 142 \\ \times\,4 \\ \hline \end{array}$$

$$\begin{array}{r} 4{,}990 \\ \times\,2 \\ \hline \end{array}$$

$$\begin{array}{r} 632 \\ \times\,7 \\ \hline \end{array}$$

$$\begin{array}{r} 37 \\ \times\,8 \\ \hline \end{array}$$

$$\begin{array}{r} 28 \\ \times\,3 \\ \hline \end{array}$$

$$\begin{array}{r} 25 \\ \times\,32 \\ \hline \end{array}$$

$$\begin{array}{r} 52 \\ \times\,4 \\ \hline \end{array}$$

$$\begin{array}{r} 63 \\ \times\,41 \\ \hline \end{array}$$

$$\begin{array}{r} 18 \\ \times\,5 \\ \hline \end{array}$$

$$\begin{array}{r} 15 \\ \times\,18 \\ \hline \end{array}$$

$$\begin{array}{r} 872 \\ \times\,6 \\ \hline \end{array}$$

$$\begin{array}{r} 562 \\ \times\,4 \\ \hline \end{array}$$

$$\begin{array}{r} 12 \\ \times\,8 \\ \hline \end{array}$$

456 9,980 800 270

459 4,424 208 5,232

534 296 2,583 2,248

568 84 90 96

Use <, >, or = to compare the fractions.

$$\frac{2}{6} \bigcirc \frac{4}{12}$$

Use <, >, or = to compare the fractions.

$$\frac{15}{45} \bigcirc \frac{5}{15}$$

Use <, >, or = to compare the fractions.

$$\frac{2}{5} \bigcirc \frac{4}{7}$$

Use <, >, or = to compare the fractions.

$$\frac{10}{12} \bigcirc \frac{9}{13}$$

Find the equivalent.

$$\frac{3}{4} = \frac{}{32}$$

Find the equivalent.

$$\frac{1}{6} = \frac{}{18}$$

Find the equivalent.

$$\frac{2}{5} = \frac{}{25}$$

Find the equivalent.

$$\frac{7}{8} = \frac{}{64}$$

$$\frac{3}{5} \times 9 =$$

$$\frac{1}{6} \times 7 =$$

$$8 \times \frac{1}{7} =$$

$$2 \times \frac{2}{5} =$$

$$8 \times \frac{1}{10} =$$

$$3 \times \frac{2}{7} =$$

$$8 \times \frac{1}{2} =$$

$$2 \times \frac{7}{9} =$$

$>$ $<$ $=$ $=$

56 10 3 24

$\dfrac{4}{5}$ $1\dfrac{1}{7}$ $1\dfrac{1}{6}$ $5\dfrac{2}{5}$

$1\dfrac{5}{9}$ 4 $\dfrac{6}{7}$ $\dfrac{4}{5}$

$\begin{array}{r} 2{,}586 \\ \times \quad 3 \\ \hline \end{array}$ © CD	$\begin{array}{r} 42 \\ \times 13 \\ \hline \end{array}$ © CD	$\begin{array}{r} 65 \\ \times 13 \\ \hline \end{array}$ © CD	$\begin{array}{r} 236 \\ \times \quad 9 \\ \hline \end{array}$ © CD
$6\overline{)24}$ © CD	$8\overline{)48}$ © CD	$9\overline{)108}$ © CD	$2\overline{)15}$ © CD
$4\overline{)124}$ © CD	$5\overline{)2{,}145}$ © CD	$5\overline{)75}$ © CD	$3\overline{)148}$ © CD
$4\overline{)1{,}218}$ © CD	$4\overline{)1{,}728}$ © CD	$6\overline{)2{,}346}$ © CD	$5\overline{)252}$ © CD

2,124	845	546	7,758
7r1	12	$\underline{6}$	4
49r1	15	429	31
50r2	391	432	304r2

$$\frac{1}{2} \times 3 =$$

$$6 \times \frac{5}{9} =$$

$$\begin{array}{r}\frac{5}{6}\\[2pt]+\ \frac{9}{6}\\\hline\end{array}$$

$$\begin{array}{r}\frac{2}{18}\\[2pt]+\ \frac{3}{18}\\\hline\end{array}$$

$$\begin{array}{r}5\frac{1}{4}\\[2pt]+\ 6\frac{3}{4}\\\hline\end{array}$$

$$\begin{array}{r}2\frac{3}{8}\\[2pt]+\ 4\frac{1}{8}\\\hline\end{array}$$

$$\begin{array}{r}\frac{2}{5}\\[2pt]+\ \frac{4}{5}\\\hline\end{array}$$

$$\begin{array}{r}\frac{2}{9}\\[2pt]+\ \frac{1}{9}\\\hline\end{array}$$

$$\begin{array}{r}4\frac{1}{8}\\[2pt]+\ 2\frac{3}{8}\\\hline\end{array}$$

$$\begin{array}{r}1\frac{1}{2}\\[2pt]+\ 8\frac{1}{2}\\\hline\end{array}$$

$$\begin{array}{r}2\frac{2}{3}\\[2pt]+\ 5\frac{2}{3}\\\hline\end{array}$$

$$\begin{array}{r}\frac{7}{12}\\[2pt]-\ \frac{5}{12}\\\hline\end{array}$$

$$\begin{array}{r}\frac{5}{7}\\[2pt]-\ \frac{3}{7}\\\hline\end{array}$$

$$\begin{array}{r}\frac{7}{8}\\[2pt]-\ \frac{5}{8}\\\hline\end{array}$$

$$\begin{array}{r}\frac{4}{6}\\[2pt]-\ \frac{1}{6}\\\hline\end{array}$$

$$\begin{array}{r}6\frac{1}{4}\\[2pt]-\ 2\frac{1}{4}\\\hline\end{array}$$

© CD

$$\frac{5}{18}$$

$$\frac{1}{3}$$

$$\frac{1}{6}$$

$$4$$

$$2\frac{1}{3}$$

$$1\frac{1}{5}$$

$$8\frac{1}{3}$$

$$\frac{1}{2}$$

$$3\frac{1}{3}$$

$$6\frac{1}{2}$$

$$10$$

$$\frac{1}{4}$$

$$1\frac{1}{2}$$

$$12$$

$$6\frac{1}{2}$$

$$\frac{2}{7}$$

$$2\frac{2}{3} - 2\frac{1}{3}$$

$$\begin{array}{r} 23 \\ \times\ 15 \\ \hline \end{array}$$

$$\begin{array}{r} 11 \\ \times\ 22 \\ \hline \end{array}$$

$$\begin{array}{r} 143 \\ \times\ 8 \\ \hline \end{array}$$

$$2\frac{1}{4} - \frac{3}{4}$$

$$\begin{array}{r} 162 \\ \times\ 3 \\ \hline \end{array}$$

$$\begin{array}{r} 9{,}632 \\ \times\ 5 \\ \hline \end{array}$$

$$\begin{array}{r} 533 \\ \times\ 4 \\ \hline \end{array}$$

$$\frac{13}{15} - \frac{12}{15}$$

$$\begin{array}{r} 152 \\ \times\ 8 \\ \hline \end{array}$$

$$\begin{array}{r} 49 \\ \times\ 83 \\ \hline \end{array}$$

$$\begin{array}{r} 16 \\ \times\ 4 \\ \hline \end{array}$$

$$4\frac{1}{8} - 2\frac{3}{8}$$

$$4\frac{7}{10} - 1\frac{4}{10}$$

$$\begin{array}{r} 65 \\ \times\ 54 \\ \hline \end{array}$$

$$\begin{array}{r} 102 \\ \times\ 6 \\ \hline \end{array}$$

$\frac{1}{3}$ $1\frac{1}{2}$ $\frac{1}{15}$ $1\frac{3}{4}$

345 486 1,216

242 48,160 4,067 3,510

1,144 2,132 64 612

Change to a fraction.

0.8

Change to a fraction.

0.99

Change to a decimal.

$\dfrac{40}{100}$

Change to a decimal.

$\dfrac{5}{100}$

Change to a fraction.

0.75

Change to a fraction.

0.30

Change to a decimal.

$\dfrac{2}{10}$

Change to a decimal.

$\dfrac{61}{100}$

Name the type of angle.

Name the type of angle.

Name the type of angle.

Name the type of angle.

Name the type of angle.

Name the type of angle.

Name the type of angle.

Name the type of angle.

0.05	0.40	$\frac{99}{100}$	$\frac{8}{10}$
0.61	0.2	$\frac{3}{10}$	$\frac{75}{100}$
obtuse	right	acute	obtuse
acute	acute	obtuse	right